EXERCISE WORKBOOK

for

Beginning AutoCAD®

2007

by

Cheryl R. Shrock

Professor
Drafting Technology
Orange Coast College, Costa Mesa, Ca.

INDUSTRIAL PRESS
New York

Industrial Press Inc.
989 Avenue of the Americas
New York, NY 10018

10 9 8 7 6 5 4 3

This book is dedicated to the teachers that continue to use my workbooks within their AutoCAD courses and the independent users that seek an easy to use instruction book to guide them through their introduction to AutoCAD.

You are the reason I write these workbooks.

Exercise Workbooks written by Cheryl R. Shrock:

Advanced AutoCAD **2000**	ISBN 0-8311-3193-4
Beginning AutoCAD **2000, 2000i & LT**	ISBN 0-8311-3194-2
Advanced AutoCAD **2000, 2000i & LT**	ISBN 0-8311-3195-0
Beginning AutoCAD **2002**	ISBN 0-8311-3196-9
Advanced AutoCAD **2002**	ISBN 0-8311-3197-7
Beginning AutoCAD **2004**	ISBN 0-8311-3198-5
Advanced AutoCAD **2004**	ISBN 0-8311-3199-3
Beginning AutoCAD **2005**	ISBN 0-8311-3200-0
Advanced AutoCAD **2005**	ISBN 0-8311-3201-9
Beginning AutoCAD **2006**	ISBN 0-8311-3213-2
Advanced AutoCAD **2006**	ISBN 0-8311-3214-0
Beginning AutoCAD **2007**	ISBN 0-8311-3302-3
Advanced AutoCAD **2007**	ISBN 0-8311-3303-1

For information about these workbooks,
visit www.industrialpress.com

For information about Cheryl Shrock's online courses,
visit www.shrockpublishing.com

Table of Contents

INTRODUCTION

About this workbook

Exercise Workbook for Beginning AutoCAD® 2007 is designed for classroom instruction or self-study. There are 30 lessons. Each lesson starts with step-by-step instructions followed by exercises designed for practicing the commands you learned within that lesson.

You may find the order of instruction in this workbook somewhat different from most textbooks. The approach I take is to familiarize you with the drawing commands first. After you are comfortable with the drawing commands, you will be taught to create your own setup drawings. This method is accomplished by supplying you with drawings "Workbook Helper" and "9A Helper". These drawings are preset and ready for you to open and use. For the first 8 lessons you should not worry about settings, **you just draw.**

I realize that not everyone will agree with this approach and if this is the case, you may want to change the order in which the lessons are learned. I have had success with this method because my students feel less intimidated and more confident. This feeling of confidence increases student retention. Learning should be fun not a headache.

The exercises in the workbook, that include printing, are designed for a Hewlett Packard 4MV printer capable of printing a 17 X 11 drawing. These exercises can be amended to match your printer or plotter specifications. To configure your printer, refer to Appendix A, "Add a Printer / Plotter. But it is important to note that you have the ability to configure a printer / plotter <u>even though your computer is not attached to it</u>. I advise you to configure the HP 4MV, to complete the lessons within the workbook, even though you will not use this printer for actual printing.

Important
How to get the drawings listed above?
The 2 files mentioned above should be downloaded from one of the following:
http://www.industrialpress.com or www.shrockpublishing.com

About the Author

Cheryl R. Shrock is a Professor and Chairperson of Computer Aided Design at Orange Coast College in Costa Mesa, California. She is also an Autodesk® registered author / publisher. Cheryl began teaching CAD in 1990. Previous to teaching, she owned and operated a commercial product and machine design business where designs were created and documented using CAD. This workbook is a combination of her teaching skills and her industry experience.

"Sharing my industry and CAD knowledge has been the most rewarding experience of my career. Students come to learn CAD in order to find employment or to upgrade their skills. Seeing them actually achieve their goals, and knowing I helped, is a real pleasure. If you read the lessons and do the exercises, I promise, you will not fail."

Cheryl R. Shrock

Configuring your system

AutoCAD ® 2007 allows you to customize it's configuration. While you are using this workbook, it is necessary for you to make some simple changes, to your configuration, so our configurations are the same. This will ensure that the commands and exercises work as expected. The following instructions will guide you through those changes.

A. First start AutoCAD®
   1. Follow the instructions on page 1-7 and 1-8. Then return to this next step B.

B. At the bottom of the screen there is a white rectangular area called the "*Command Line*". Type: ___*Options*___ then press the **<enter>** key. (not case sensitive)

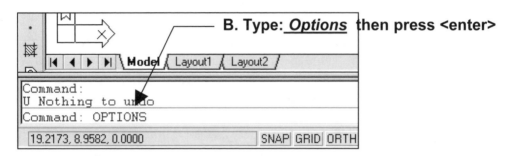

C. Select the ***Display*** tab and change the settings on your screen to match the dialog box below. Pay special attention to the settings with an ellipse around it.

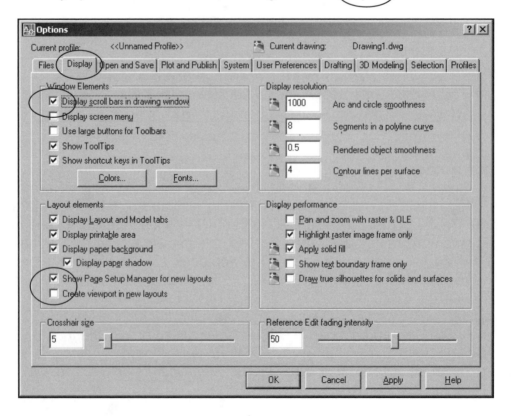

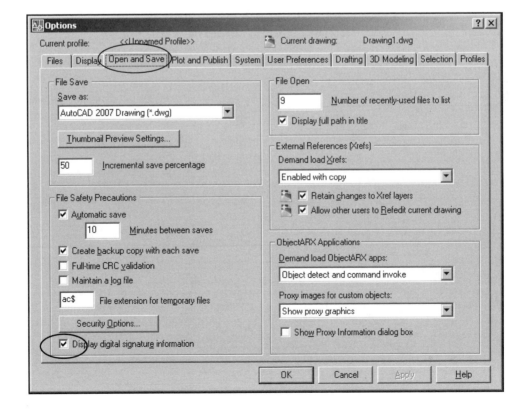

2007 LT

E. Select the **Open and Save** tab and change the settings on your screen to match the dialog box below.

2007

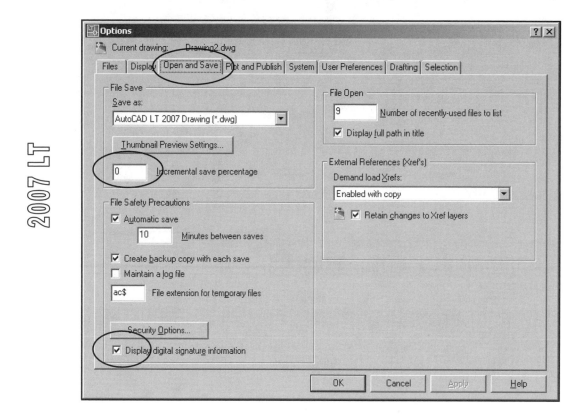

F. Select the **Plot and Publish** tab and change the settings on your screen to match the dialog box below.

IMPORTANT: Add this printer.
See Appendix A for instructions
(Don't worry, it is not difficult)

Yours will be different.

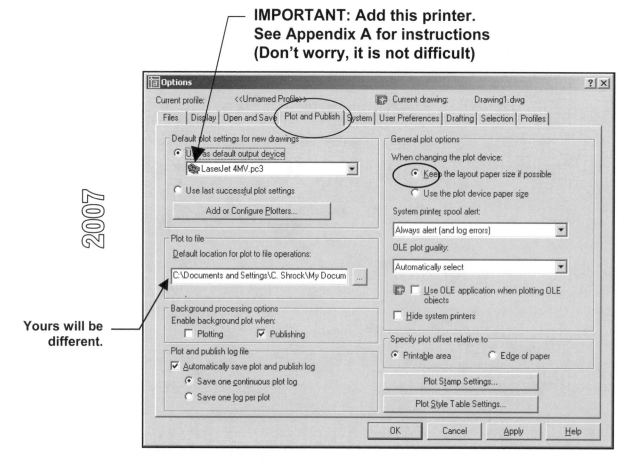

IMPORTANT: Add this printer.
See Appendix A for instructions
(Don't worry, it is not difficult)

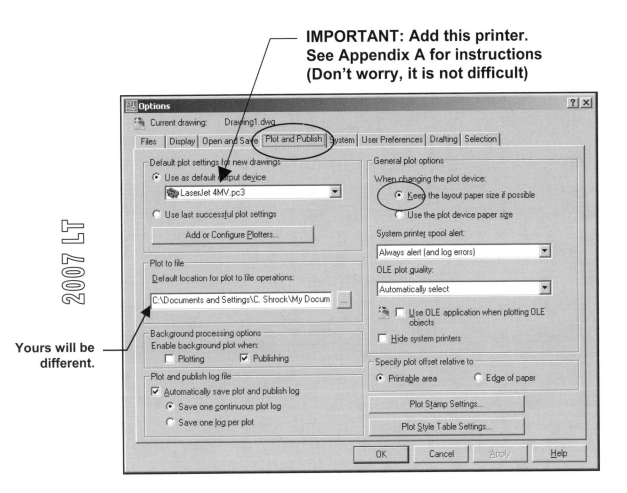

2007 LT

Yours will be different.

G. Select the **System** tab and change the settings on your screen to match the dialog box below.

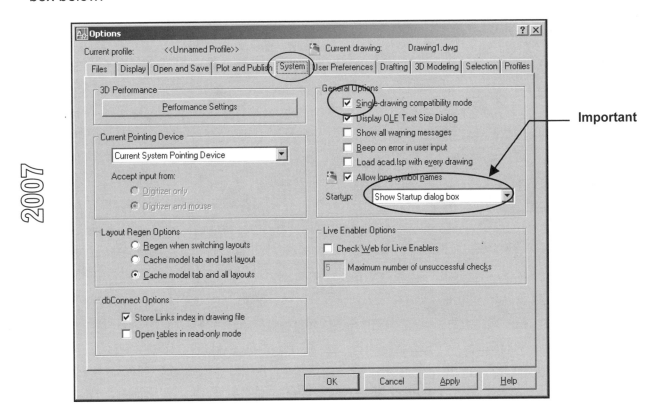

2007

Important

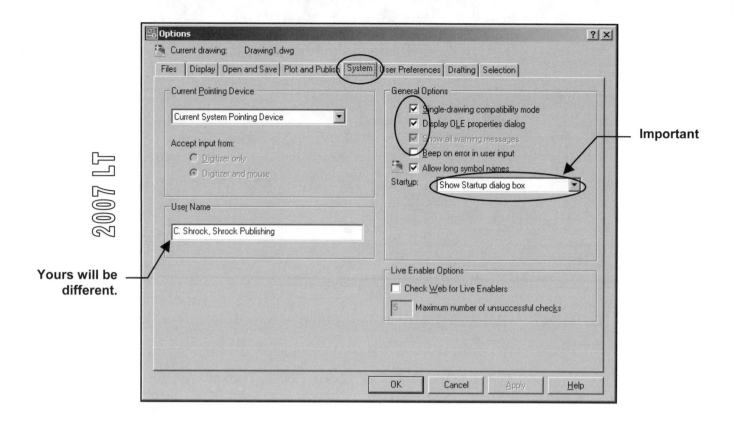

2007 LT

Important

Yours will be different.

H. Select the *User Preferences* tab and change the settings on your screen to match the dialog box below.

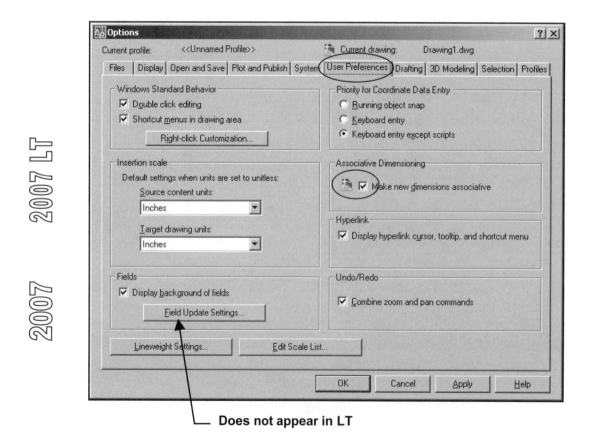

2007 LT

2007

Does not appear in LT

I. Select the **Right-click Customization..** box and change the settings on your screen to match the dialog box below.

Select "Right-click customization" button

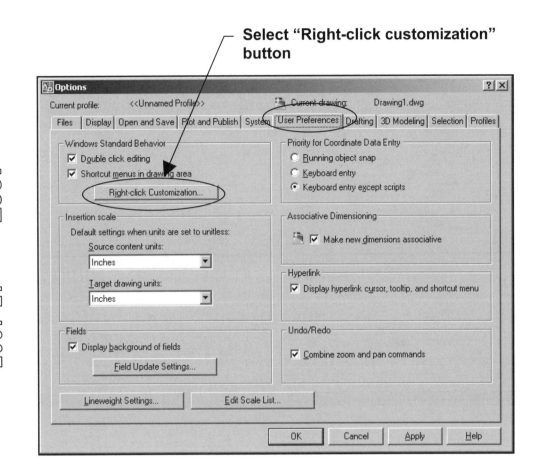

J. Select button before going on to the next tab.

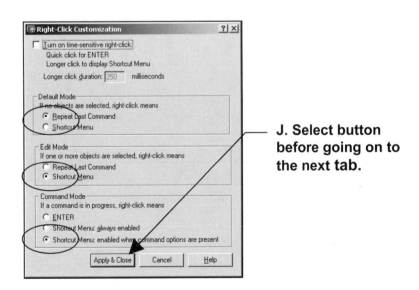

J. Select the **Apply & Close** button, shown above, before going on to the next tab.

K. Select the **Drafting** tab and change the settings on your screen to match the dialog box below.

Options

Current profile: <<Unnamed Profile>>   Current drawing: Drawing1.dwg

| Files | Display | Open and Save | Plot and Publish | System | User Preferences | Drafting | 3D Modeling | Selection | Profiles |

AutoSnap Settings
- ☑ Marker
- ☑ Magnet
- ☑ Display AutoSnap tooltip
- ☑ Display AutoSnap aperture box

Colors...

AutoSnap Marker Size

Object Snap Options
- ☑ Ignore hatch objects
- ☐ Replace Z value with current elevation
- ☑ Ignore negative Z object snaps for Dynamic UCS

AutoTrack Settings
- ☑ Display polar tracking vector
- ☑ Display full-screen tracking vector
- ☑ Display AutoTrack tooltip

Alignment Point Acquisition
- ◉ Automatic
- ○ Shift to acquire

Aperture Size

Drafting Tooltip Settings...

Lights Glyph Settings...

Cameras Glyph Settings...

[OK]   [Cancel]   [Apply]   [Help]

2007

Options

Current drawing: Drawing1.dwg

| Files | Display | Open and Save | Plot and Publish | System | User Preferences | Drafting | Selection |

AutoSnap Settings
- ☑ Marker
- ☑ Magnet
- ☑ Display AutoSnap tooltip
- ☑ Display AutoSnap aperture box

Colors...

AutoSnap Marker Size

Object Snap Options
- ☑ Ignore hatch objects

AutoTrack Settings
- ☑ Display polar tracking vector
- ☑ Display full-screen tracking vector
- ☑ Display AutoTrack tooltip

Alignment Point Acquisition
- ◉ Automatic
- ○ Shift to acquire

Aperture Size

Drafting Tooltip Settings...

[OK]   [Cancel]   [Apply]   [Help]

2007 LT

L. Select the **Selection** tab and change the settings on your screen to match the dialog box below.

3D tab does not appear in LT

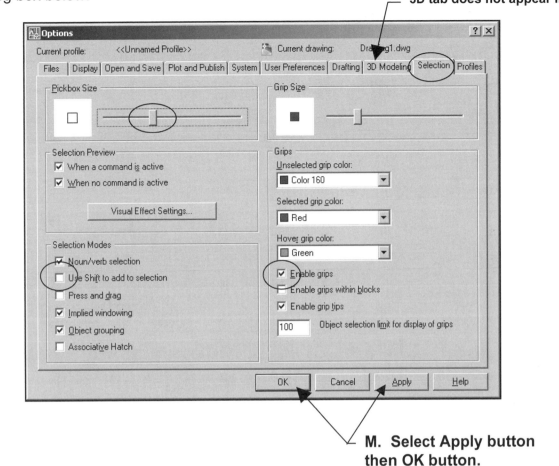

2007

2007 LT

M. Select Apply button then OK button.

M. Select the **Apply** button then the **OK** button.

Customizing your Wheel Mouse

A Wheel mouse has two or more buttons and a small wheel between the two topside buttons. The default functions for the two top buttons are as follows:

Left Hand button is for **input**

Right Hand button is for **Enter** or the **shortcut menu**.

You will learn more about this later. But for now follow the instructions below.

Using a Wheel Mouse with AutoCAD®

To get the most out of your Wheel Mouse set the **MBUTTONPAN** setting to **"1"** as follows:

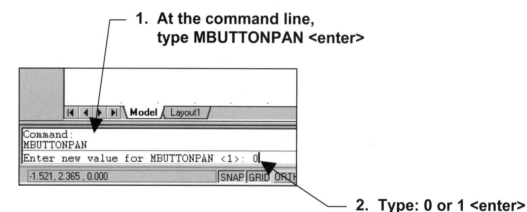

1. **At the command line, type MBUTTONPAN <enter>**

2. **Type: 0 or 1 <enter>**

After you understand the function of the "Mbuttonpan" variable, you can decide whether you prefer the setting "0" or "1" as described below.

MBUTTONPAN setting 0:

ZOOM   Rotate the wheel forward to zoom in
                       Rotate the wheel backward to zoom out

OBJECT   Object Snap menu will appear when you press the wheel
SNAP

MBUTTONPAN setting 1: (Factory setting)

ZOOM   Rotate the wheel forward to zoom in
                       Rotate the wheel backward to zoom out

ZOOM
EXTENTS   Double click the wheel

PAN   Press the wheel and drag

LEARNING OBJECTIVES

After completing this lesson, you will be able to:

1. Understand basic computer terms.
2. Understand what CAD means.
3. Determine what computer to purchase.
4. Know the system requirements for AutoCAD.
5. Start AutoCAD four ways.
6. Use AutoCAD's Help system

LESSON 1

Part 1. UNDERSTANDING COMPUTERS

A BRIEF HISTORY OF COMPUTERS AND SOFTWARE.
The first computers were developed in the 1950s, shortly after the transistor was invented. In the mid 1960s General Motors, Boeing and IBM began developing CAD programs, but the development was slowed by the high cost of computer hardware and programming.

In 1971, Ted Hoff developed the first microprocessor. All circuitry of the central processing unit (CPU) was now on one chip. This started the era of the personal computer (PC). In the 1980s, additional improvements to the microprocessor changed the mainframe computers to powerful desktop models.

Of course, computer software was advancing along with the computer hardware. CAD started as a simple drafting tool and has now evolved into a powerful design tool. CAD has progressed from two-dimensional (2-D) to three-dimensional (3-D), to surface modeling and to solid modeling with animation. Each generation has become more powerful and more user friendly.

HARDWARE

Microprocessor
The complex procedure that transforms raw input data into useful information for output is called "processing". The **processor** is the "brain" of the computer. The processor interprets and carries out instructions. In personal computers the processor is a single chip plugged into a circuit board. This chip is called a **microprocessor**.

Central Processing Unit (CPU)
The CPU is the term used for the computer's processor. The CPU contains the intelligence of the machine. It is where the calculations and decisions are made.

Memory (RAM)
Your CPU needs memory to hold pieces of information while it works. While this information remains in memory, the CPU can access it directly. This memory is called **random access memory (RAM)**. RAM holds information only while the power is on. When you turn off or reset the computer, the information disappears.
The more RAM a computer has, the quicker it works and the more it can do.
The most common unit of measurement for computer memory is the **byte**. A **byte** can be described as the amount of memory it takes to store a single character. A **kilobyte (KB)** equals 1,024 bytes. A **Megabyte (MB)** equals 1,024 kilobytes, or 1,048,576 bytes. So a computer with 64 MB of memory actually has (64 X 1,048,576) 67,108,864 bytes. This is equal to approximately 1024 pages of information.

Input / Output devices
Input devices accept data and instructions from the user. The most common input devices are the keyboard, mouse and scanner. Output devices return processed data back to the user. The most common output devices are the monitor, printer and speakers.

Storage

The purpose of storage is to hold data that the computer isn't using. When you need to work with a set of data, the computer retrieves the data from storage and puts it into memory. When it no longer needs the data, it puts it back into storage. There are 2 advantages to storage. First, there is more room in storage and second, storage retains its contents when the computer is turned off. Storage devices include: Hard disks (inside your computer), floppy disks (3-1/2) , zip disks, CDR/W, etc.

SOFTWARE

Operating Systems

When you turn on the computer, it goes through several steps to prepare itself for use.

The first step is a self-test. This involves:
   a. Identifying the devices attached to it (such as the monitor, mouse and printer).
   b. Counts the amount of memory available.
   c. Checks to see if the memory is functioning properly.

The second step is searching for a special program called the Operating System. When the computer finds the operating system, it loads it into memory (remember RAM). The operating system enables the computer to:
   a. Communicate with you.
   b. Use devices such as the disk drives, keyboard and monitor.

The operating system is now ready to accept commands from you. The operating system continues to run until the computer is turned off. Examples of operating systems are: Windows NT, ME, 2000, XP, OS/2, Unix and more.

Note: 1. Apple / Macintosh computers have their own operating system.
2. AutoCAD 2007 will not work with Windows 98, NT, ME or Apple / Macintosh.

Application software

The operating system is basically for the computer. The Application Software is for the user. Application Software is designed to do a specific task.
There are basically four major categories:
Business, Utility, Personal, and Entertainment.

Business application software would be desktop publishing, spreadsheet programs, database software and graphics. *AutoCAD is a "graphics" business application software.*

Utility application software helps you maintain your computer. You would use a utility program to recover an accidentally deleted file, improve the efficiency of your computer and help you move, copy or delete files. *Norton Utilities is an example of a "utility application" software.*

Personal application software is basically what it sounds like. This software is designed for your personal needs, such as: balancing your checkbook, making an address book, creating a calendar and many more tasks.

Entertainment application software are video games, puzzles, flight simulators and even educational programs.

COMPUTER SIZES AND CAPABILITIES

Computers are divided into five catagories:
Supercomputer, Mainframe, Minicomputer, Workstation and Personal Computer.

Supercomputers are the <u>most powerful</u>. These computers process huge amounts of information very quickly. For example, scientists build models of complex processes and simulate the processes on a supercomputer.

Mainframe computers are the <u>largest</u>. These computers are designed to handle tremendous amounts of input, output and storage. For example, the government uses mainframe computers to handle the records for Social Security.

Minicomputers are <u>smaller than mainframe</u> computers, but bigger than personal computers. They do not handle as much as the mainframe computers, but they are less expensive. A company that needs the features of a mainframe, but can't afford such a large computer, may choose a minicomputer.

Workstations <u>resemble a personal computer</u> but are much <u>more powerful</u>. Their internal construction is different than a PC. Workstations use a different CPU design called "reduced instruction set computing" (RISC), which makes the instructions process faster. Scientists and engineers generally use workstations, using the UNIX operating system.

Personal computers (PC), originally named microcomputers, are <u>small computers</u> that usually reside on a desktop. This category would include Laptops.

What is a Clone?
In 1981, IBM called its first microcomputer the IBM PC. Many companies copied this design and they functioned just like the original. These copies were called "clones" or "compatibles". The term PC is now used to describe this family of computers.
Note: the Apple/Macintosh computer is neither an IBM or a compatible. It should <u>not</u> be called a PC.

Part 2. What is CAD?

Computer Aided Design (CAD) is simply, **design** and **drafting** with the aid of a computer. <u>Design</u> is creating a real product from an idea. <u>Drafting</u> is the production of the drawings that are used to document a design. CAD can be used to create 2D or 3D computer models. A CAD drawing is a file that consists of numeric data in binary form that will be saved onto a disk.

Why should you use CAD?
Traditional drafting is repetitious and can be inaccurate. It may be faster to create a simple "rough" sketch by hand but larger more complex drawings with repetitive operations are drawn more efficiently using CAD.

Why use AutoCAD?
AutoCAD is a computer aided design software developed by Autodesk Inc.
AutoCAD was first introduced in 1982. By the year 2000, it is estimated that there were over 4 million AutoCAD users worldwide.

What this means to you is that many employers are in need of AutoCAD operators.
In addition, learning AutoCAD will give you the basics for learning other CAD packages because many commands, terms and concepts are used universally.

Part 3. Buying your first computer

Buying your first computer is not an easy task. Here are a few tips:

1. Make a list of tasks for which you will use your computer
    a. Select the software for those tasks.
    b. Select the computer that will run that software.

2. Talk to other computer owners.
    a. Listen to their good and bad experiences.

3. Educate yourself.
    a. Go to your local library and spend an evening reading through the computer magazines. Most are written with the novice in mind.

4. Decide how much you can afford.
    a. Remember, you can always upgrade or add components later.

5. Find the right company to buy from. A great deal can turn into a bad investment if you can't get help when you need it.
    a. Ask about their customer service and support.
    b. How long is the warranty?

Part 4. AutoCAD 2007 system requirements

Operating system:
> Microsoft Windows XP Professional SP 1 or 2
> Microsoft Windows XP Home SP 1 or 2
> Microsoft Windows XP Tablet PC SP 2
> Microsoft Windows 2000 SP 3 or 4 (SP4 recommended)

Browser
> Microsoft Internet Explorer 6.0 SP 1or later

RAM and Hard Disk Space
> 512 MB of RAM minimum
> 750MB of hard disk space
> 64MB of swap space (Recommended)
> 100MB free disk space in system folder (Recommended)

Hardware (required)
> Pentium III or later with 800 MHz processor or better
> Mouse, Track Ball or other pointing device
> 1024 x 768 video graphics display with true color
> CD-ROM drive for initial installation only

Hardware (optional)
> Printer or Plotter
> Serial or Parallel port (for peripheral devices)
> Sound card with speakers
> Disk drive (3-1/2 floppy, Zip or CDW) for saving files.

STARTING AutoCAD

To Start AutoCAD:

1. Select the **START** button, located in the lower left corner of the screen.

2. Select **ALL PROGRAMS**

3. Select **Autodesk**

4. Select **AutoCAD 2007** or **AutoCAD LT 2007**

5. Select **AutoCAD 2007** or **AutoCAD LT 2007**

Note: For AutoCAD 2007 LT:
The New Features dialog box will appear. Select "Maybe Later" then OK.

For AutoCAD 2007:
The dialog box shown below should appear.

6. Select "**AutoCAD Classic**". (**3D Modeling** will be discussed in the Advanced Workbook)

7. Select "**OK**" button.

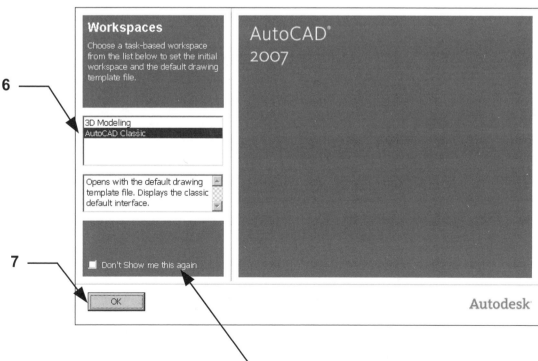

Note: If the above dialog box does not appear, someone may have checked the "Don't show me this again" box. If so, don't be alarmed. You may select "AutoCAD Classic" using another method shown on the next page.

Your screen should now appear as shown below. Notice the words "**AutoCAD Classic**" appears in the <u>workspace box</u> in the upper left corner of the screen. If the words "**3D Modeling**" appear in this box click on the down arrow and select "**AutoCAD Classic**".

<u>Workspace box</u> should display "**AutoCAD Classic**".
Note: LT: The Workspace Box is floating, not docked.

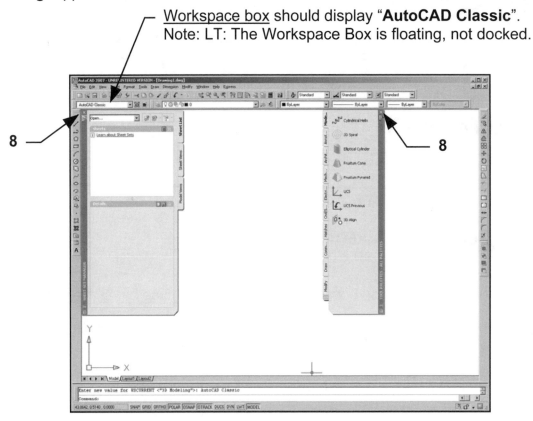

8. Now close the **Sheet Manager** (LT has "Quick Help") and the **Tools Palettes** by selecting the **"X"** located in the upper left or right corner of each.

Your screen should now appear as shown below.

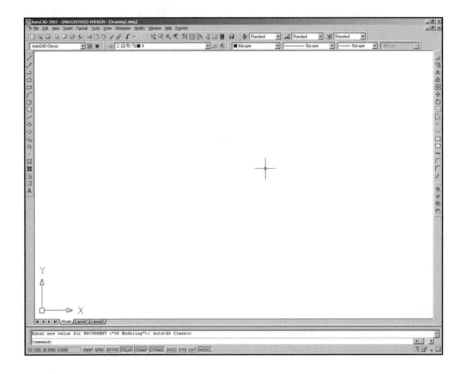

STARTING AutoCAD continued......

Select **FILE / NEW**

> **NOTE:** If one of the dialog boxes shown below does not appear refer to page Intro-5
> item G to change your setting for "Startup".

*I prefer to use the dialog boxes for students new to AutoCAD. After you become an
"expert" you may prefer to disable this option.*

Notice the four buttons located in the upper left corner of this dialog box. Each button
provides a different way to start a drawing. A brief description of each is listed below.

Open a Drawing
This option is only active when you first enter AutoCAD.
Normally you will use **File / Open.** Refer to page 2-18.
Select a drawing from the list shown of the most recently
opened drawings or select the "Browse" button to search for
more drawing files. After you select the file desired, select
the OK button. The file selected will appear on your screen.

Start from Scratch
Begin a new drawing from scratch. Starting from scratch
means all settings are preset by AutoCAD.
You must select the measurement system on which to base
your new drawing; Imperial or Metric.

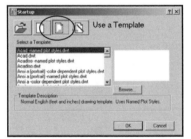

Use a Template
Choose a previously created template. You can choose one
of the templates supplied with AutoCAD or create your own.
Note:
We will be creating a Template in Lesson 2.

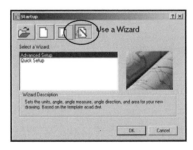

Use a Wizard
Start a new drawing using either the "Quick" or "Advanced"
setup wizard. The wizard sets the units, angle, angle
measurement, angle direction and area for your new
drawing. (You will learn all of these settings in Lesson 9.)

Using AutoCAD's HELP system

This workbook was created to make your AutoCAD learning experience fun and uncomplicated.   To accomplish this I do not include every little detail about each command. I teach you the "meat and potatoes" of AutoCAD. If you would like to learn more about a command or system variable try AutoCAD's HELP system.

HOW TO OPEN THE HELP SYSTEM

Method 1.
1. Start a command.
2. Press the F1 key.

Method 2.
1. Select HELP menu at the top of the screen.
2. Select HELP from the drop down menu.

Method 3.
1. Press F1
2. Click on the "AutoCAD 2007 Help" button on the task bar.

USING THE HELP SYSTEM

The Contents tab   Organizes by topic like a table of contents in a book.

The Index tab   Alphabetical listing of topics. Type the first few letters of the word. As you type, the list jumps to the closest match.

The Search tab   Find keywords. Type a word in the text box.

The Ask Me tab   Ask a question and hopefully get an answer. Type a question or phrase and press <enter>.

Concepts   The overall description.

Procedures   How to do it.

Commands   Related commands.

Quick Help on the Info Palette

"Quick Help" provides a brief explanation about an AutoCAD command. You may choose to search for the command alphabetically or display a list of procedures <u>as you work</u>.

How to display the Info Palette
Select "Info Palette" from the Help menu at the top of your screen or hold down the "Ctrl key" and press 5.

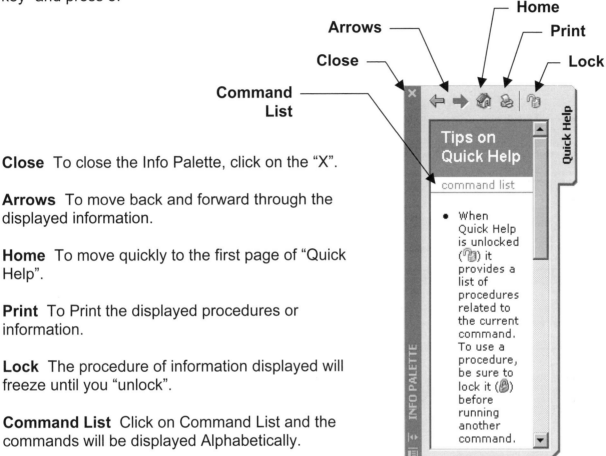

Close To close the Info Palette, click on the "X".

Arrows To move back and forward through the displayed information.

Home To move quickly to the first page of "Quick Help".

Print To Print the displayed procedures or information.

Lock The procedure of information displayed will freeze until you "unlock".

Command List Click on Command List and the commands will be displayed Alphabetically.

Command List

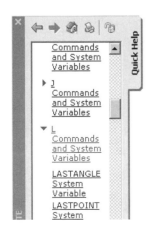

Procedure displayed as you draw

NOTES:

LEARNING OBJECTIVES

After completing this lesson, you will be able to:

1. Create a template.
2. Understand the AutoCAD Window.
3. Understand the use of the function keys.
4. Select commands using the Pull-down Menu Bar, Toolbars or by typing at the Command Line.
5. Recognize a dialog box.
6. Open, Close and Move a toolbar.
7. Draw, Erase and Select Lines.
8. Clear the screen.
9. Save a drawing.
10. Open an existing drawing.
11. Exit AutoCAD.

LESSON 2

CREATE A TEMPLATE

The first item on the learning agenda is how to create a template file from the **"Workbook Helper.dwg"**. Go to the website: *http://www.industrialpress.com*, select AutoCAD under "Books by Top Subject", to download the files for the 2007 workbook and save them to your "Desktop".

Now we will create a template. This will be a very easy task.

1. Start AutoCAD, if you haven't already. (Refer to page 1-7)

2. Select **File / Open**

3. Select the **Desktop** directory and locate the downloaded the files mentioned above.

4. Select the file "**Workbook Helper.dwg**" and then "**Open**" button.

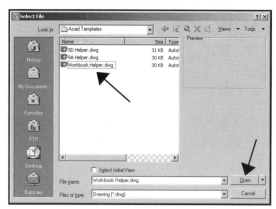

Notice the 3 letter extension for a "drawing" file is ".dwg".

5. Select **"File / Save As..."**

6. Select the "**Files of type:**" down arrow ▾ to display different saving formats. Select "**AutoCAD Drawing Template (*.dwt)**".

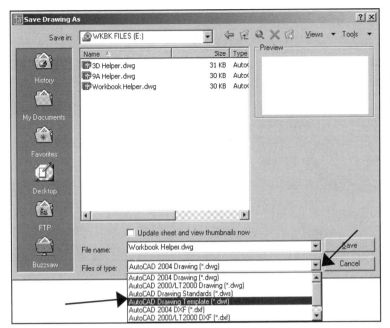

Notice the 3 letter extension for Template is ".dwt".

A list of all the AutoCAD templates will appear. (Note: Your list may be different)

7. Type the new name "**1Workbook Helper**" in the "**File name:**" box and then select the "**Save**" button.

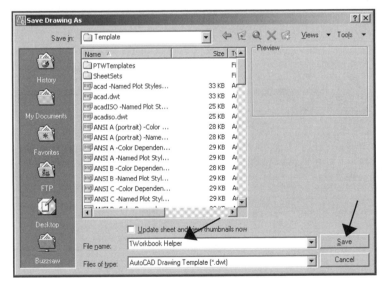

Note: The "1" before the name will place the file at the top of the list.
AutoCAD displays numerical first and then alphabetical.

Notice it was not necessary to type the extension .dwt because "Files of type" was previously selected.

8. Type a description and the select the "**OK**" button.

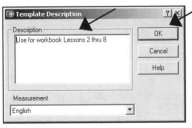

Now you have a template to use for lessons 2 through 8. At the beginning of each of the exercises you will be instructed to open this template.
Using a template as a master setup drawing is good CAD management.

OPENING A TEMPLATE

The template that you created on the previous page will be used for lessons 2 through 8. It will appear as a blank screen, but there are many variables that have been preset. This will allow you to start drawing immediately. You will learn how to set those variables before you complete this workbook, but for now you will concentrate on learning the AutoCAD commands and hopefully have some fun.

To Open a Template follow the steps below.

1. Select **FILE / NEW**.

2. Select the **Use a Template** box (third from the left).

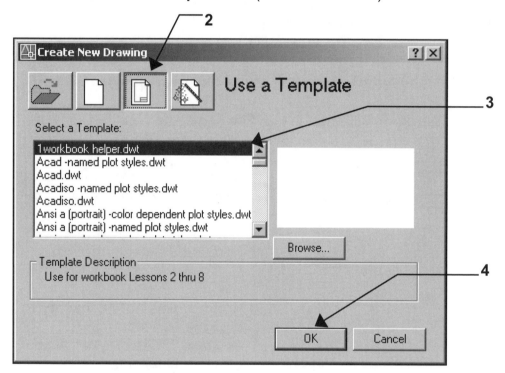

3. Select **1workbook helper.dwt** from the list of templates.

   (NOTE: If you do not have this template, refer to page 2-2.)

4. Select the **OK** button.

NOTE: If you find that you have more than one drawing open, it is important that you have configured your AutoCAD software to only allow one drawing open at one time. It will be less confusing for now. When you progress to the Advanced workbook, you will configure AutoCAD for multiple open drawings. But for now, refer to Intro-5G for "Single drawing compatibility mode" setting under "General Options". Check this option box.

GETTING FAMILIAR WITH THE AUTOCAD WINDOW

Before you can start drawing you need to get familiar with the AutoCAD window. In the following lessons, I will be referring to all of the areas described below. So it is important for you to understand each of them. But remember, this page will always be here for you to refer back.

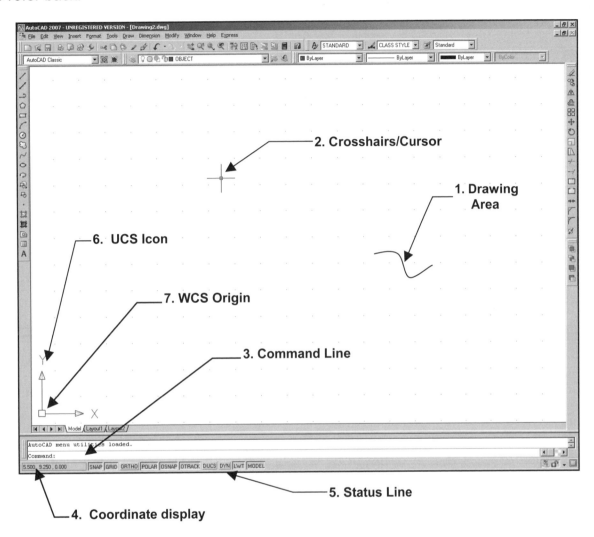

1. DRAWING AREA
   Location: The large area in the center of the screen.
   This is where you will draw. This area represents a piece of paper.
   The color of this area can be changed using Tools / Option / Display / Color.
   The default color for 2007 is Black.

2. CROSSHAIRS / CURSOR
   Location: Can be anywhere in the Drawing Area.
   The movement of the pointing device such as a mouse controls the movement of the cursor. You will use the cursor to locate points, make selections and draw objects.
   The size can be changed using Tools / Options / Display / Crosshair Size.

3. COMMAND LINE
Location: The three lines at the bottom of the screen.
This is where you enter commands and Autocad will prompt you to input information.
More on this later.

4. COORDINATE DISPLAY (F6)
Location: Lower left corner

In the **Absolute mode (coords = 1)**: displays the location of the crosshairs / cursor in reference to the Origin. The first number represents the horizontal movement (Xaxis), the second number represents the vertical movement (Yaxis) and the third number is the Zaxis which is used for 3D.

In the **Relative Polar mode (coords = 2)**: displays the distance and angle of the cursor from the last point entered. (Distance<Angle)

5. STATUS LINE
Location: Below the Command Line.
Displays your current settings. These settings can be turned on and off by clicking on the word (Snap, Grid, Ortho, etc.) or by pressing the function keys, F1, F2, etc.
A brief description of each listed below. More detailed information later.

[SNAP] (F9)
Increment Snap controls the movement of the cursor. If it is off, the cursor will move smoothly. If it is ON, the cursor will jump in an incremental movement.
The increment spacing can be changed at any time using **Tools / Drafting Settings / Snap and Grid**. The default spacing is .250.

[GRID] (F7)
The grid (dots) is merely a visual "drawing aid". The default spacing is 1 unit.
You may change the grid spacing at any time using: **Tools / Drafting Settings / Snap and Grid.**

[ORTHO] (F8)
When Ortho is ON, cursor movement is restricted to horizontal or vertical. When Ortho is OFF, the cursor moves freely.

[POLAR] (F10)
POLAR TRACKING creates "Alignment Paths" at specified angles.

[OSNAP] (F3)
RUNNING OBJECT SNAP (More detailed information on page 4-4)
Specific Object Snaps can be set to stay active until you turn them off.

[OTRACK] (F11)
OBJECT SNAP TRACKING
Creates "Alignment Paths" at precise positions using object snap locations.

[DUCS] (F11) (Not available in LT)
DYNAMIC USER COORDINATE SYSTEM
Used for 3D. Refer to Advanced Workbook.

[DYN]
DYNAMIC INPUT
Displays the Dynamic Command Input by the cursor. This can be very distracting for students new to AutoCAD. **For now I would like this option OFF. Just click on the DYN button so it is not depressed.** (More information on page 11-7)

[LWT]
LINEWEIGHT. Displays the width assigned to each object. (More information on page 9-7)

[MODEL]
Switches your drawing between paperspace and modelspace.
(More information in Lesson 26)

6. **UCS ICON (User Coordinate System)**
   Location: Lower left corner of the screen. The UCS icon indicates the location of the Origin. The UCS icon appearance can be changed using: **View / Display / Icon / Properties**.

7. **WCS ORIGIN (World Coordinate System)**
   The location where the X, Y and Z axis intersect. 0,0,0
   (Don't worry about this now. We will talk more in Lesson 9)

FUNCTION KEYS

F1	Help	Explanations of Commands. (Refer to page 1-9)
F2	Flipscreen	Toggles from Text Screen to Graphics Screen.
F3	Osnap	Toggles Osnap On and Off.
F4	Tablet	Toggles the Tablet On and Off.
F5	Isoplane	Changes the Isoplane from Top to Right to Left.
F6	DUCS	Toggles Dynamic UCS On / Off.
F7	Grid	Toggles the Grid On or Off.
F8	Ortho	Toggles Ortho On or Off.
F9	Snap	Toggles Increment Snap On or Off.
F10	Polar	Toggles Polar Tracking On or Off.
F11	Otrack	Toggles Object Snap Tracking On and Off.
F12	Dynamic Input	Toggles Dynamic Input On and Off.

SPECIAL KEY FUNCTIONS

Escape Key Cancels the current operation.

Enter Key Ends a command, or will repeat the previous command if the command line is blank.

Space Bar Same as the Enter Key, except when entering text.

PULL-DOWN "MENU BAR"

(1) The pull-down "MENU BAR" is located at the top of the screen.

By selecting any of the words in the MENU BAR, a **(2) Pull-down menu** appears.
If you select a word from the pull-down menu that has an **(3) Arrow ➤** , a
(4) Sub Menu will appear.   (Example : Draw / Circle)
If you select a word with **(5) Ellipsis ...** , a dialog box will appear.
(Example: Draw / Boundary...)

(1) Pull-down "MENU BAR"

(2) Pull-down Menu

3. Arrow

4. Sub Menu

5. Ellipsis "..."

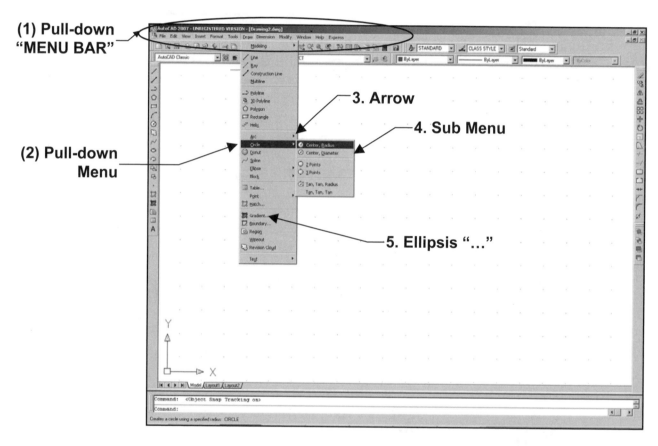

DIALOG BOX

Many commands have **multiple options** and require
you to make selections. These commands will display
a dialog box. Dialog boxes, such as the *Hatch* dialog
box shown here, make selecting and setting options
easy.

2-8

TOOLBARS

AutoCAD provides several toolbars to access frequently used commands.

The **(1) Standard**, **(2) Styles** **(3) Workspace** **(4) Layers** **(5) Properties**, **(6) Draw, and (7) Modify** toolbars are displayed by default.

Toolbars contain *icon buttons (8).* These icon buttons can be selected to Draw or Edit objects and manage files.

If you place the pointer on any icon and wait a second, a **tool tip (9)** will appear and a **help message (10)** will appear at the bottom of the screen.

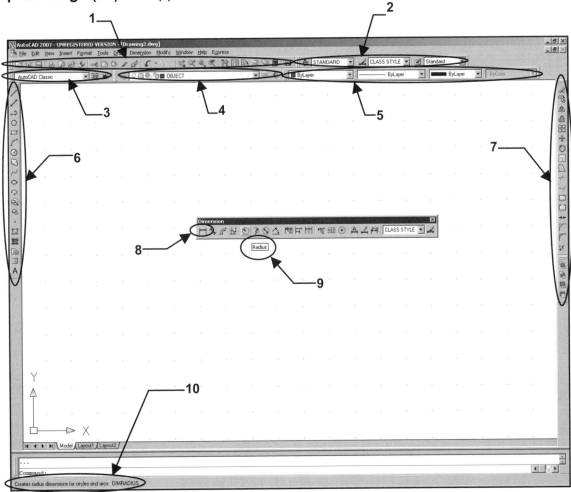

Toolbars can be **"Floated"** or **"Docked".**

Floating toolbars (11) move freely in the drawing area and can be resized.

To move, place the pointer on the toolbar title then hold the left mouse button down, drag to the new location and release the mouse button.

To resize, place the pointer on the right or bottom edge of the toolbar. When the pointer changes to a double ended arrow, hold the left mouse button down and drag. When desired size is achieved, release the mouse button.

Docked toolbars (12) are locked into place along the top, bottom or sides of the AutoCad Window.

To dock, place the pointer on the toolbar title, hold the left mouse button down and drag to the top, bottom, or either side of the AutoCAD window. When the outline of the toolbar appears, release the mouse button.

HOW TO OPEN TOOLBARS

Many other toolbars are available.

1. Place the cursor in the gray area to the left or right side of the screen.

2. Press the right mouse button and the list shown will appear.

3. Select a toolbar by placing the cursor on the name and press the left mouse button. The check indicates that the toolbar is already open.

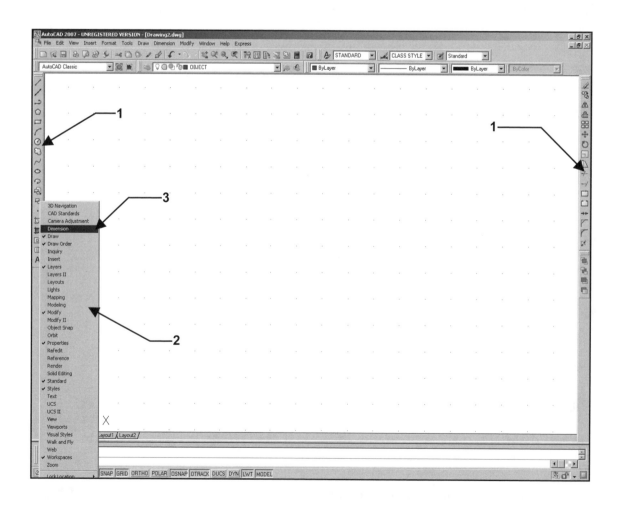

PALETTES

A Palette organizes tools or operations for easy access. Three examples are shown below. Some are for direct use and some can be customized.

AutoCAD has more than a dozen palettes. AutoCAD LT has approximately half dozen.

Palettes may be resized and moved to any location on the screen. They can float or be docked. The Auto-Hide function allows you to collapse the palette when the cursor is away from the palette. When you move the cursor over the Title Bar the Palette will reappear.

Note: You will learn much more about Palettes through out this workbook. This is merely an introduction.

Refer to the Advanced Workbook for more detailed instructions for customizing.

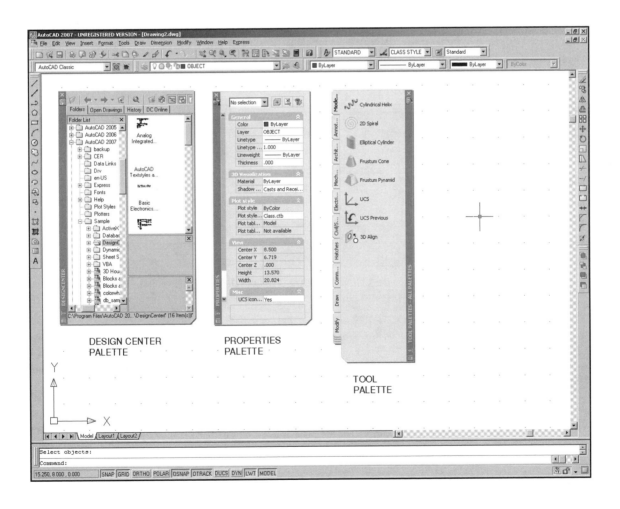

METHODS OF ENTERING COMMANDS

AutoCAD has 3 different methods of entering commands. All 3 methods will accomplish the same end result. AutoCAD allows you to use the method you prefer. The following are descriptions of all 3 methods and an example of how each one would be used to start a command, such as the **Circle** command.

1. **Pull down Menu** (page 2-8) **(Select Draw / Circle / Center Radius)**
   a. Move the cursor to the Menu Bar (1).
   b. Click on a Menu header such as "Draw".
   c. Slide the cursor down the list of commands and left click to select.

2. **Tool Bars** (page 2-9) **(Select the Circle icon** 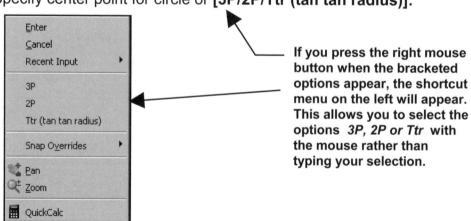 **from the Draw toolbar)**
   Move the cursor to an icon on a toolbar and press the left mouse button to select.

3. **Keyboard (Type C and <enter>.)**
   Type the command on the command line.
   Note: Dynamic Input is also a keyboard method. But you should have this option temporarily turned off. To turn DYN on or off press F12 or use the DYN button on the Status line. (Dynamic Input will be discussed in lesson 11.)

What is a SHORTCUT menu?
In addition to the methods listed above, AutoCAD has shortcut menus. Shortcut Menus give you quick access to command options. Shortcut Menus are only available when brackets [ ] enclose the options on the command line. (Example below)
To activate a Shortcut Menu, press the right mouse button when the bracketed options appear on the command line.

Example:

> Select: Draw / Circle / Center, Radius
> _circle Specify center point for circle or **[3P/2P/Ttr (tan tan radius)]:**

If you press the right mouse button when the bracketed options appear, the shortcut menu on the left will appear. This allows you to select the options *3P, 2P or Ttr* with the mouse rather than typing your selection.

Selecting the **Shortcut menus** is the AutoCAD's **"Heads Up"** drawing method within AutoCAD. "Heads Up" means always looking up at the screen and not down at the keyboard. Using this method should improve your efficiency and productivity. You will learn more about AutoCAD's Heads Up method using Dynamic Input in lesson 11.

DRAWING LINES

A **_LINE_** can be _one segment_ or a _series of connected segments_. Each segment is an individual object.

One segment   **Series of connected segments**

Start the Line command by using one of the following methods:

Type = L <enter>
PULLDOWN MENU = DRAW / LINE
TOOLBAR = DRAW

Lines are drawn by specifying the locations for each endpoint.
Move the cursor to the location of the **"first"** endpoint (1) then press the left mouse button. Move the cursor again to the **"next"** endpoint (2) and press the left mouse button. Continue locating **"next"** endpoints until you want to stop.

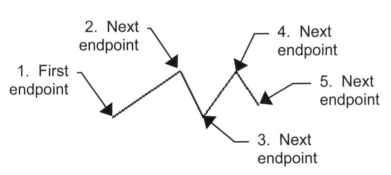

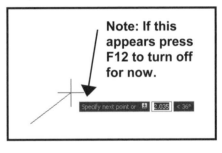

Note: If this appears press F12 to turn off for now.

There are 3 ways to **Stop** drawing a line: **1**. Press <enter> key. **2**. Press <Space Bar> or **3**. Press the right mouse button then select enter from the short cut menu.

To draw a perfectly **Horizontal** or **Vertical** line select the **ORTHO** mode by clicking on the **ORTHO button** on the **Status Bar** or pressing **F8**.
**Note: Ortho can be temporarily turned off while drawing by holding down the "Shift key". Release the "Shift key to resume.**

If you have drawn two or more line segments, the _endpoint_ of the _last line segment_ can be connected automatically to the first endpoint using the **CLOSE** option.

To use this option, draw two or more line segments, then type **C** <enter>.

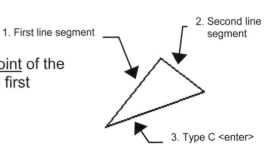

1. First line segment
2. Second line segment
3. Type C <enter>

ERASE

There are 3 methods to **erase** (delete) objects from the drawing.
You decide which one you prefer to use. They all work equally well.
(See "Methods of Selecting Objects" on page 2-15)

METHOD 1.
Select the Erase command first and then select the objects.

1. Start the Erase command by using one of the following:

> **TYPING = E <enter>**
> **PULLDOWN = MODIFY / ERASE**
> **TOOLBAR = MODIFY**

2. Select objects: *pick one or more objects*
    Select objects: *press <enter> and the objects will disappear*
    (Note: AutoCAD will continue to prompt you to select more objects until you press <enter> to make it stop.)

METHOD 2.
Select the Objects first and then the <u>Erase</u> command from the shortcut menu.

1. Select the object(s) to be erased.
2. Press the right mouse button.
3. Select **<u>Erase</u>** from the short-cut menu.

METHOD 3.
Select the Objects first and then the <u>Delete</u> key

1. Select the object(s) to be erased.
2. Press the <u>Delete</u> key.

NOTE: Very important
If you want the erased objects to return, press **U <enter>** or **Ctrl + Z** or
the **Undo arrow icon.**

This will **"Undo"** the effects of the last command.

Read more about the **Undo** and **Redo** commands on page 3-6.

METHODS OF SELECTING OBJECTS

Most AutoCAD commands prompt you to "select objects". This means select the objects that you want the command to effect.

There are 2 methods. **Method 1. <u>Pick</u>**, is very easy and should be used if you have only 1 or 2 objects to select. **Method 2. <u>Window</u>**, is a little more difficult but once mastered it is extremely helpful and time saving. Practice the examples shown below.

<u>Method 1. PICK</u> :

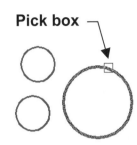

Pick box

When the command line prompt reads, "Select Objects", place the cursor (pick box) on top of the object and click the left mouse button.
The object about to be selected will highlight.
This appearance change is called "Rollover Highlighting".
This gives you a selection preview of which object will be selected.

<u>Method 2. WINDOW</u>: <u>C</u>rossing and <u>W</u>indow

<u>Crossing</u>:

Place your cursor in the area <u>up</u> and to the <u>right</u> of the objects that you wish to select (**P1**) and press the left mouse button. Then move the cursor down and to the left of the objects (**P2**) and press the left mouse button again.
(Note: The window will be *green* and outer line is *dashed*.)
<u>Only</u> the objects that this window crosses or completely enclosed will be selected.

In the example on the right, all 3 circles have been selected.
(The 2 small circles are ***completely enclosed*** and the large circle is ***crossed*** by the window.)

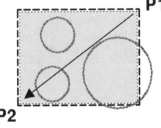

<u>Window</u>:

Place your cursor in the area <u>up</u> and to the <u>left</u> of the objects that you wish to select (**P1**) and press the left mouse button. Then move the cursor <u>down</u> and to the <u>right</u> of the objects (**P2**) and press the left mouse button.
(Note: The window will be *blue* and outer line is *solid*.)
<u>Only</u> the objects that this window completely enclosed will be selected.

In the example on the right, only 2 circles have been selected.
(The large circle is ***not*** completely enclosed.)

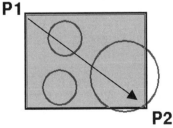

Note: If the windows described above do not show up on your screen, it means that your "implied windowing" is turned off. Select <u>Tools / Options / Selection tab</u>. In the section "<u>Selection Modes</u>" on the left, place a check mark in the "<u>Implied windowing</u>" box. Also, the colors and the opacity of the windows may be customized using the "Visual Effect Settings".

STARTING A NEW DRAWING

1.   Start the command using one of the following methods:

TYPING: **NEW <enter> or press CTRL + N**
PULLDOWN: **FILE / NEW**
TOOLBAR: **STANDARD**

The Dialog box shown below should appear.

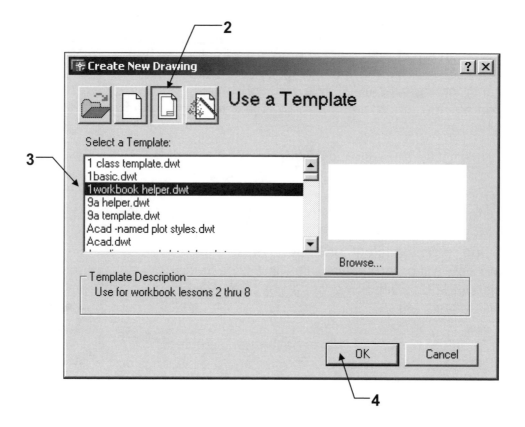

2.   Select the **Use a Template** box (third from the left).

3.   Select **1workbook helper.dwt** from the list of templates.
     (NOTE: If you do not have this template, refer to page 2-2)

4.   Select the **OK** button (bottom right).

NOTE: It is important that you have configured your AutoCAD software to only allow one drawing open. It will be less confusing for now. When you progress to the Advanced workbook, you will configure AutoCAD for multiple open drawings. Refer to Intro-5 item G for "Single drawing compatibility mode" setting.

SAVING A DRAWING

After you have completed a drawing, it is very important to save it. Learning how to save a drawing correctly is almost more important than making the drawing. If you can't save correctly, you will lose the drawing and hours of work.

*There are 2 commands for saving a drawing: **Save** and **Save As**. I prefer to use **Save As**. The **Save As** command always pauses to allow you to choose where you want to store the file and what name to assign to the file. This may seem like a small thing, but it has saved me many times from saving a drawing on top of another drawing by mistake. The **Save** command will automatically save the file either back to where you retrieved it or where you last saved a previous drawing. Neither may be the correct destination. So play it safe, use **Save As** for now.*

1. Start the command by using one of the following methods:

TYPING:	**SAVEAS <enter>**
PULLDOWN:	**FILE / SAVEAS**
TOOLBAR:	**No icon for <u>Save As</u>, only for <u>Save (Don't use it for now)</u>**

This Dialog box should appear:

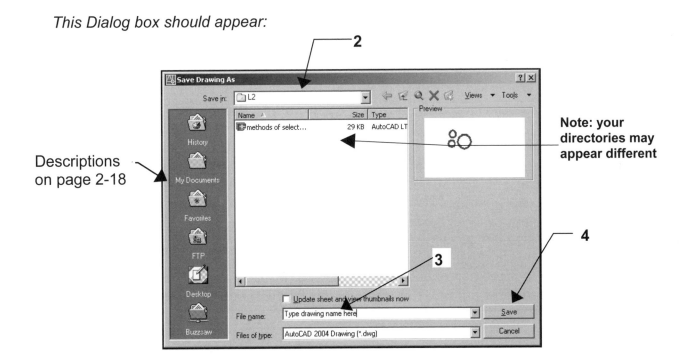

Descriptions on page 2-18

Note: your directories may appear different

2. Select the appropriate drive and directory from the **"SAVE IN"** box. (This is where your drawing will be saved)

3. Type the new drawing file name in the **"FILE NAME"** box.

4. Select the **"SAVE"** button.

BACK UP FILES

When you save a drawing file, Autocad creates a file with a **.dwg** extension.
For example, if you save a drawing as **12b**, Autocad saves it as **12b.dwg**. The next time you save that same drawing, Autocad replaces the old with the new and renames the old version **12b.bak**. The old version is now a back up file.
(Only 1 backup file is stored.)

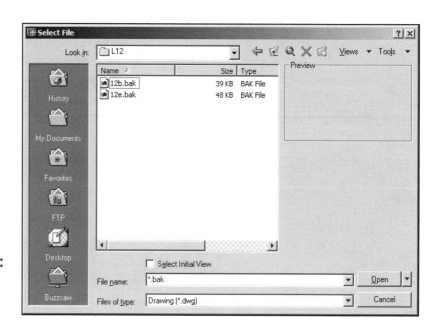

How to open a back up file:
You can't open a .bak file.
It must first be renamed with a .dwg file extension.

How to view the list of back up files:
Select File / Open
Type "*.bak" in the "file name" box and <enter>. A list of the backup (.bak) files, within the chosen directory, will appear.

How to rename a back up file:
Right click on the file name. Select "Rename". Change the .bak extension to .dwg and press <enter>.

The following is information only. We will not be using these in this workbook.

History: Displays shortcuts to the files most recently accessed from the dialog box. Note: The shortcuts remain until you remove them. This list can get very long. To delete the shortcuts go to: Windows / Application Data / Autodesk / AutoCAD / Recent / Save Drawing As.

Desktop: Displays the contents of your desktop.

My Docs: Displays the contents of the Personal or My Documents folder for the current user profile. The name of this location depends on your operating system version.

Favorites: Displays the contents of the Favorites folder for the current user profile.

Buzzsaw: Provides access to projects hosted by Buzzsaw.com—a business-to-business marketplace for the building design and construction industry.

RedSpark: Provides access to projects hosted by RedSpark—a business-to-business marketplace for the manufacturing industry.

FTP: Displays the FTP sites that are available for browsing in the standard file selection dialog box. (FTP means: File Transfer Protocol)

OPENING AN EXISTING DRAWING FILE

1. Start the command by using one of the following methods.

TYPING: **OPEN <enter>** or press **CTRL + O**
PULLDOWN: **FILE / OPEN**
TOOLBAR: **STANDARD**

The Dialog box shown below should appear.

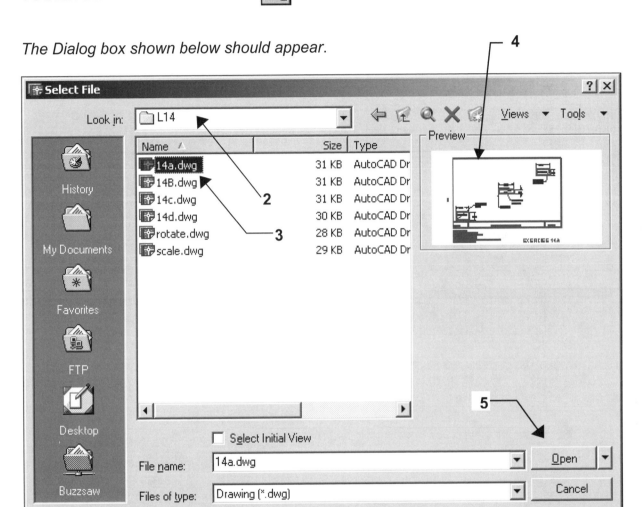

2.   Select the **Drive and Directory** from the **"LOOK IN"** Box.
3.   Select the drawing file from the list. (You may double click on the file name to automatically open the drawing)
4.   The Preview window displays a "Thumbnail Preview Image".
5.   Select the **OPEN** button.

NOTE: It is important that you have configured your AutoCAD software to only allow one drawing to be open. It will be less confusing for now. When you progress to the Advanced workbook, you will configure AutoCAD for multiple open drawings. Refer to Intro-5 item G for "Single drawing compatibility mode" setting.

EXITING AUTOCAD

1. Start the command by using one of the following methods:

TYPING: **EXIT <enter> or   QUIT<enter>**
PULLDOWN: **FILE / EXIT (Safest method)**
TOOLBAR: **NONE**

If any changes have been made to the drawing since the last save, the warning box below will appear asking if you want to **SAVE THE CHANGES**?

Select **YES, NO** or **CANCEL**.

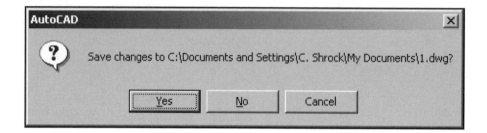

EXERCISE 2A

INSTRUCTIONS:

1. Start a **New** file (refer to 2-16) and select **1workbook helper.dwt**.
2. **Draw** the objects below using:
   - **LINE** command
   - Ortho (f8) **ON** for **Horizontal** and **Vertical** lines
   - Ortho (f8) **OFF** for lines drawn on an **Angle.**
   - Increment Snap (f9) **ON**
   - Osnap (f3) **OFF**
3. **Save** this drawing using:
   - File / Save as / **EX2A**

Note: Please confirm that Dynamic Input is "Off" by releasing the "**DYN**" button on the Status bar. (Dynamic Input will be discussed in lesson 11.)
Also, if your lines appear too thick, release the **LWT** button on the Status bar.

EXERCISE 2B

INSTRUCTIONS:

1. Using drawing **EX2A**, **ERASE** the missing lines.
2. **Save** this drawing using:
   File / Save as / **EX2B**

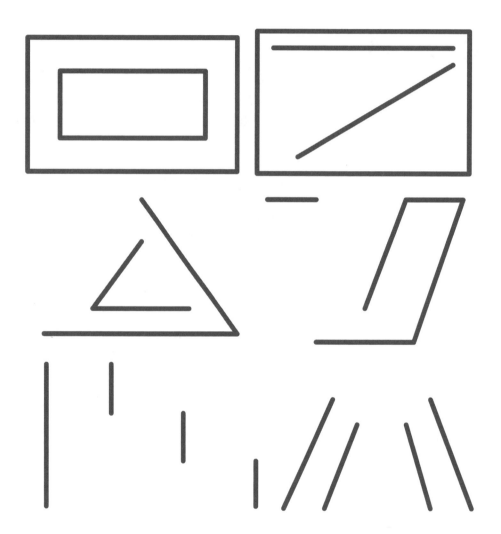

EXERCISE 2C

INSTRUCTIONS:

1. Start a **New** file and select **1workbook helper.dwt**.
2. **Draw** the objects below using:
   Draw / Line
   Ortho (F8) **ON** for **Horizontal** and **Vertical** lines
   Ortho (F8) **OFF** for lines drawn on an **Angle.**
   Increment Snap (F9) **ON**
   Osnap (F3) **OFF**
3. **Save** this drawing using:
   File / Save as / **EX2C**

EXERCISE 2D

INSTRUCTIONS:

1. Start a **New** file and select **1workbook helper.dwt**.
2. **Draw** the objects below using:
   Draw / Line
   Ortho (f8) **ON** for **Horizontal** and **Vertical** lines
   Ortho (f8) **OFF** for lines drawn on an **Angle.**
   Increment Snap (f9) **ON**
   Osnap (f3) **OFF**
3. **Save** this drawing using:
   File / Save as / **EX2D**

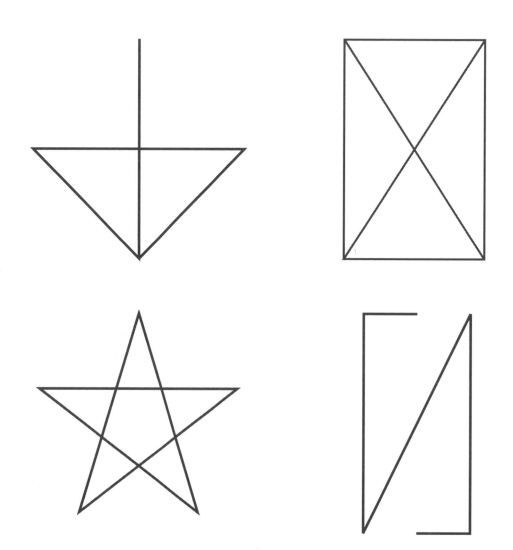

LEARNING OBJECTIVES

After completing this lesson, you will be able to:

1. Create a Circle using 6 different methods.
2. Create a Rectangle with width, chamfers, fillets and rotation.
3. Set Grids and Increment Snap using the Drafting Settings option.
4. Change current Layers.

LESSON 3

CIRCLE

*There are 6 options to create a circle. The default option is "**Center, radius**".*
(Probably because that is the most common method of creating a circle.)
*We will try the "**Center, radius**" option first.*

1. Start the **Circle** command by using one of the following:
   TYPING = C <enter>
   PULLDOWN = DRAW / CIRCLE / Center, Radius
   TOOLBAR = DRAW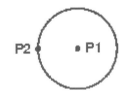

2. The following will appear on the command line:
   Command: _circle Specify center point for circle or [3P/2P/Ttr (tan tan radius)]:

3. Locate the center point for the circle by moving the cursor to the desired location in the drawing area and press the left mouse button.

4. Now move the cursor away from the center point and you should see a circle forming.

5. When it is approximately the size desired, press the left mouse button, or if you want the exact size, type the radius and then press <enter>.

Note: To use one of the other methods described below, first select the Circle command, then press the right mouse button. A "short cut" menu will appear. Select the method desired by placing the cursor on the option and pressing the left mouse button. Or you can type 3P or 2P or T, then press <enter>. (The short cut menu is simple and more efficient.)

Center, Radius: (Default option)
   1. Specify the center (P1) location.
   2. Specify the Radius (P2).

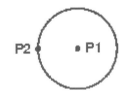

Center, Diameter:
   1. Specify the center (P1) location.
   2. Select the Diameter option using the shortcut menu or type "D" <enter>.
   3. Specify the Diameter (P2).

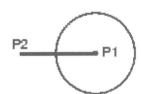

2 Points:
1. Select the 2 point option using the short cut menu or type 2P <enter>.
2. Specify the 2 points (P1 and P2) that will determine the Diameter .

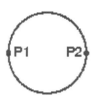

3 Points:
1. Select the 3 Point option using the short cut menu or type 3P <enter>.
2. Specify the 3 points (P1, P2 and P3) on the circumference.
   The Circle will pass through all three points.

Tangent, Tangent, Radius:
1. Select the Tangent, Tangent, Radius option using the short cut menu or type T <enter>.

2. Select two objects (P1 and P2) for the Circle to be tangent to by placing the cursor on the object and pressing the left mouse button

3. Specify the radius.

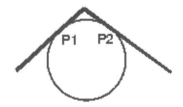

Tangent, Tangent, Tangent:
1. Select the Tangent, Tangent, Tangent option using the pull down menu. Note: May only be selected using the pull down menu. It is not available in the short cut menu or the command line.

2. Specify three objects (P1, P2 and P3) for the Circle to be tangent to by placing the cursor on the object and pressing the left mouse button. (The diameter will be calculated by the computer.)

RECTANGLE

A Rectangle is a closed rectangular shape. It is one object not 4 lines.
You can specify the length, width, area, and rotation parameters.
You can also control the type of corners on the rectangle—fillet, chamfer, or square and the width of the Line.

First, let's start with a simple Rectangle using the mouse to select the corners.

1. Start the **RECTANGLE** command by using one of the following:

    TYPING = REC <enter>
    PULLDOWN = DRAW / RECTANGLE
    TOOLBAR = DRAW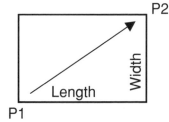

2. The following will appear on the command line:

Command: _rectang
Specify first corner point or [Chamfer/Elevation/Fillet/Thickness/Width]:

3. Specify the location of the first corner by moving the cursor to a location (**P1**) and then press the left mouse button.

    The following will appear on the command line:

Specify other corner point or [Area / Dimensions / Rotation]:

4. Specify the location of the **diagonal** corner (**P2**) by moving the cursor diagonally from the first corner (**P1**) and pressing the left mouse button.

    OR

    Type **D** <enter> (or press right mouse button and select "Dimensions" from the short cut menu)
    Specify length for rectangles <0.000>: *Type length <enter>.*
    Specify width for rectangles <0.000>: *Type width <enter>.*
    Specify other corner point or [Dimension]: *move the cursor up, down, right or left to specify where you want the second corner relative to the first corner and then press <enter> or press left mouse button.*

PARAMETERS – Rotation and Area

ROTATION - You may select the desired rotation angle <u>after</u> you place the first corner and <u>before</u> you place the second corner. The base point is the first corner. <u>Note: All new rectangles within the drawing will also be rotated unless you reset the rotation.</u>

AREA – You may define the size of the rectangle by inputting the Area and the length or width. Example: If you select Area and length, AutoCAD calculates the width.

OPTIONS:

You may also preset the rectangle corners to angled or rounded and adjust the line width using the Chamfer, Fillet and Width options.

CHAMFER

A chamfer is an angled corner. The Chamfer option automatically draws all 4 corners with chamfers (all the same size). You must specify the distance for each side of the corner as distance 1 and distance 2.

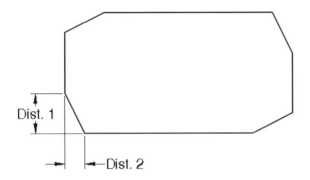

FILLET

A fillet is a rounded corner. The fillet option automatically draws all 4 corners with fillets (all the same size). You must specify the radius for the rounded corners.

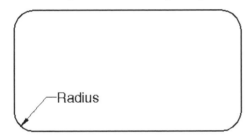

WIDTH

Sets the width of the rectangle lines. (Note: Do not confuse this with the Length and Width. This makes the lines appear to have width.)

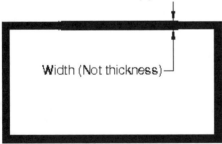

Note: If you set the Chamfer, Fillet or Width to a value greater than "0", any new rectangles will be affected until you reset the option to "0".

ELEVATION and THICKNESS: Used in 3D only. Refer to the Advanced workbook.

UNDO and REDO

The **UNDO** command allows you to undo <u>previous commands</u>. For example, if you erase an object by mistake, you can UNDO the previous "erase" command and the object will reappear. So don't panic if you do something wrong. Just use the UNDO command to remove previous commands.

Note:
You may UNDO commands used during a work session until you close the drawing.

How to use the "Undo" command.

1. Start a new drawing.
2. Draw a line, circle and a rectangle.

Your drawing should look approximately like this.

3. Erase the Circle and the Rectangle.

(*The Circle and the Rectangle disappear.*)

4. Select the UNDO down arrow and click on the ERASE command from the list of commands.

You have now deleted the ERASE command operation. As a result the erased objects reappear.

How to use the Redo command:

Select the REDO down arrow and you will see that the ERASE command moved to the REDO list. If you would like to bring the ERASE operation back, select the ERASE command from the list. The Circle and Rectangle will disappear again.

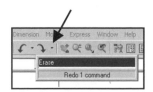

More information about the Undo and Redo commands in the Advanced Workbook.

GRID and INCREMENT SNAP

GRID is a rectangular pattern of dots in the drawing area. The grid helps you align objects and visualize the distances between them. The Grid is not plotted. Changing the "Grid X and Y spacing", within the Drafting Settings dialog box, sets the Grid dot spacing.

INCREMENT SNAP controls the movement of the cursor. If it is **OFF** the cursor will move smoothly. If it is **ON**, the cursor will jump in an *incremental* movement. This incremental movement is set by changing the **"Snap X and Y spacing"** within the Drafting Settings dialog box.

The **DRAFTING SETTINGS** dialog box allows you to set the **INCREMENT SNAP** and **GRID** spacing. You may change the Grid Spacing and Increment Snap at anytime while creating a drawing. The settings are drawing aids to help you visualize the size of the drawing and control the movement of the cursor.

1. Select **DRAFTING SETTINGS** by using one of the following:

    TYPING = DS <enter>
    PULL-DOWN = TOOLS / DRAFTING SETTINGS
    TOOLBAR = NONE

2. The dialog box shown below will appear.

3. Select the **"Snap and Grid"** tab.

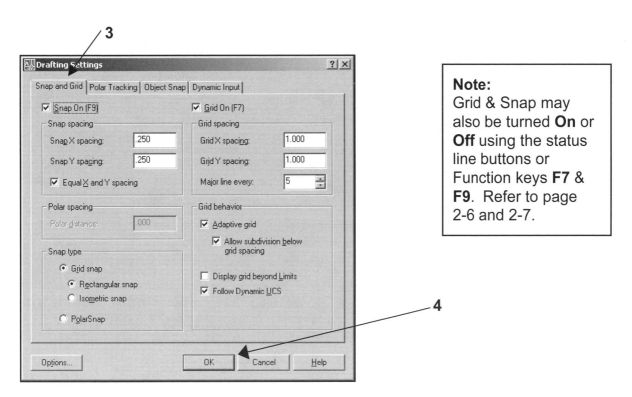

Note:
Grid & Snap may also be turned **On** or **Off** using the status line buttons or Function keys **F7** & **F9**. Refer to page 2-6 and 2-7.

4. Make your changes and select the **OK** button to save them.
    If you select the **CANCEL** button, your changes will **not** be saved.

LAYERS

A **LAYER** is like a transparency. Have you ever used an overhead light projector? Remember those transparencies that are laid on top of the light projector? You could stack multiple sheets but the projected image would have the appearance of one document. Layers are basically the same. Multiple layers can be used within one drawing.

The example, on the right, shows 3 layers.
One for annotations (text), one for dimensions
and one for objects.

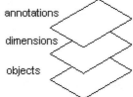

It is good "drawing management" to draw related objects on the same layer. For example, in an architectural drawing, you could have the walls of a floor plan on one layer and the Electrical and Plumbing on two other layers. These layers can then be Thawed (ON) or Frozen (OFF) independently. If a layer is Frozen, it is not visible. When you Thaw the layer it becomes visible again. This will allow you to view or make plots with specific layers visible or invisible.
(You will learn more about layers in lesson 26)

SELECTING A LAYER - Method 1. (Method 2 on the next page)

1. Display the LAYER CONTROL DROP-DOWN LIST below by clicking on the down arrow. (▼)

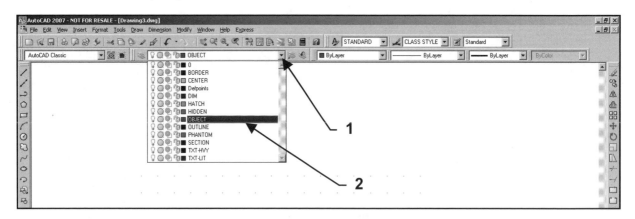

2. Click on the Layer Name you wish to select. The Layer selected will become the Current layer and the drop-down list will disappear. The Current layer means that the next object drawn will reside on this layer and will have the same color, linetype and lineweight. These are called Properties.

SELECTING A LAYER - Method 2.

1. Select the Layer command using one of the following:

   TYPE = LA <enter>
   PULLDOWN = FORMAT / LAYER
   TOOLBAR = OBJECT PROPERTIES

2. The "Layer Properties Manager" dialog box, shown below, will appear.
3. First select a layer by Clicking on its name.
4. Select the **CURRENT** button. (The green check mark)
5. Then select the **OK** button.

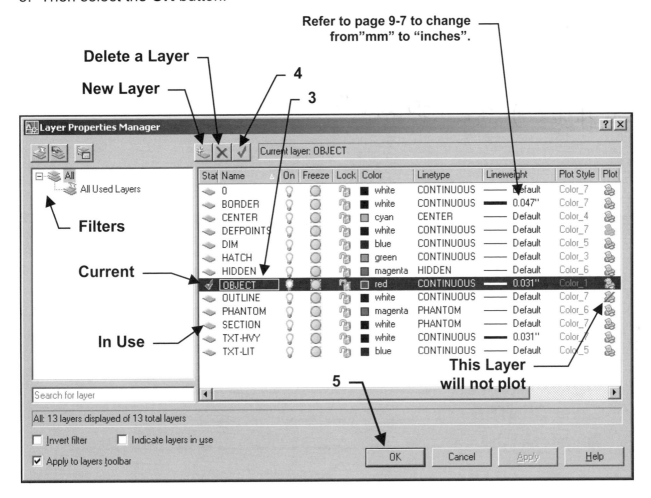

The layer you have just selected is now the **CURRENT** layer. This means that the next object drawn will reside on this layer and will have the same color, linetype and lineweight. These are called Properties.

How to delete a layer: Left click on the layer name, right click, select "Delete Layer" from the options menu. Note: You can't delete the "current" layer or a layer in use.

Layers will be discussed further in Lesson 26.

EXERCISE 3A

INSTRUCTIONS:

1. Start a **New** file and select **1workbook helper.dwt**
2. **Draw** the **LINES** below using:
   Draw / Line
   Ortho (F8) **ON** (to help you draw horizontal lines)
   Increment Snap (F9) **ON**
3. **Change** to the appropriate **layer** before drawing each line.
4. **Save** this drawing using:
   File / Save as / **EX3A**

9. dt (dekxt) size 0.25 angle '0'

Layer HIDDEN   –

Layer OBJECT   ─────────────────────

Layer PHANTOM   ─── – – ─── – – ───

Layer SECTION   ─── – – ─── – – ───

Layer TXT-HVY   ─────────────────────

Layer TXT-LIT   ─────────────────────

Layer DIM   ─────────────────────

Layer CENTER   ── – ─── – ─── – ──

Layer HATCH   ─────────────────────

EXERCISE 3B

INSTRUCTIONS:

1. Start a **New** file and select **1workbook helper.dwt**
2. Change the **GRID SPACING** to .40 and **SNAP** to .20
   using: **TOOLS / DRAFTING SETTINGS**
3. Draw the objects below, use the **layers** indicated.
4. **Save** this drawing using:
   File / Save as / **EX3B**

LAYER = OBJECT

LAYER = OBJECT

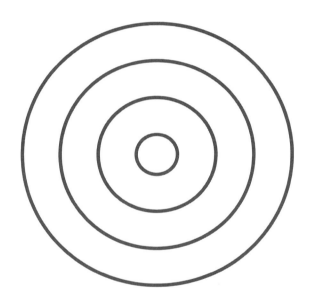

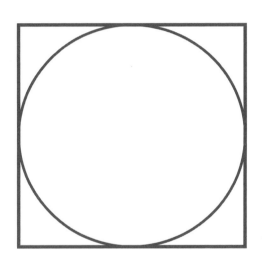

LAYER = HIDDEN

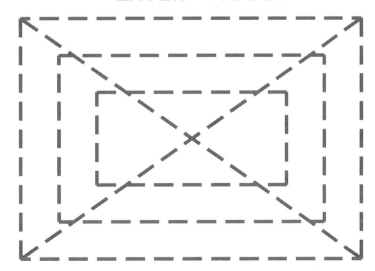

EXERCISE 3C

INSTRUCTIONS:

1. Start a **New** file and select **1workbook helper.dwt**.
2. Draw the **RECTANGLES** below using the options:
   DIMENSION, CHAMFER, FILLET, WIDTH and ROTATION
3. **Save** this drawing as: **EX3C**

Dimensions: Length = 3   Width = 2
Chamfer = .50

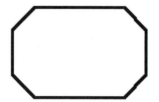

Dimensions: Length = 3   Width = 2
Fillet = Radius .75

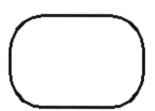

Dimensions: Length = 3   Width = 2
Rotation = 45
Chamfer = 0
Fillet = 0

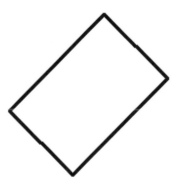

Dimensions: Length = 3   Width = 2
Width = .200
Rotation = 0
Chamfer = 0
Fillet = 0

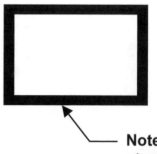

Note: Width straddles the rectangle line

EXERCISE 3D

INSTRUCTIONS:

1. Start a **New** file and select **1workbook helper.dwt.**
2. Draw the house below using at least 4 different layers.
3. You can change the **GRID** and **INCREMENT SNAP** settings to whatever you like.
4. You decide when to turn Ortho and Snap On or Off.

   Have some fun with this one!

5. Save this drawing as: **EX3D**

NOTES:

LEARNING OBJECTIVES

After completing this lesson, you will be able to:

1. Understand the function of Object Snap.
2. Use 7 Object Snap modes.
3. Operate the Running Snap function.
4. Toggle the Running Snap function On and Off.
5. Use the Zoom options to view the drawing.
6. Understand the basic concept of Setting up your drawing.
7. Change the drawing paper size.
8. Select the Units of Measurement to draw with.

LESSON 4

OBJECT SNAP

In Lesson 2 you learned about <u>Increment Snap.</u>   Increment Snap enables the cursor to move in an incremental movement. So you could say your cursor is "snapping to increments" preset by you.

Now you will learn about <u>Object Snap</u>. If Increment Snap snaps to increments, what do you think Object Snap snaps to? That's right; "objects". Object snap enables you to snap to "objects" in very specific and accurate locations on the objects. For example: The endpoint of a line or the center of a circle.

Selecting an Object Snap option using the Toolbar: Right click on any of the toolbars, then select the "Object Snap" toolbar from the list.

Selecting an Object Snap option using a Popup Menu:
<u>Method 1</u>: While holding down the shift key, press the right mouse button and the Object Snap menu will appear.

<u>Method 2</u>: Press the wheel and the Object Snap menu will appear.
(Note: The command **"Mbuttonpan"** must be set to **0** for this option to function. *Refer to Intro-10)*

OBJECT SNAP OPTIONS: *(Note: Refer to Lesson 5 for more Object Snap selections.)*

ENDpoint — Snaps to the closest endpoint of a Line, Arc or polygon segment. Place the cursor on the object close to the end.

MIDpoint — Snaps to the middle of a Line, Arc or Polygon segment. Place the cursor anywhere on the object.

INTersection — Snaps to the intersections of any two objects. Place the Pick box directly on top of the intersection or select one object and then the other and Autocad will locate the intersection.

CENter — Snaps to the center of an Arc, Circle or Donut. Place the cursor on the object, or the approximate center location.

QUAdrant — Snaps to a 12:00, 3:00, 6:00 or 9:00 o'clock location on a circle. Place the cursor on the circle near the desired quadrant location.

PERpendicular — Snaps to a point perpendicular to the object selected. Place the cursor anywhere on the object.

TANgent — Calculates the tangent point of an Arc or Circle. Place the cursor on the object as near as possible to the expected tangent point.

How to use OBJECT SNAP

The following is an example of attaching a line segment to previously drawn vertical lines. The new line will start from the upper endpoint **(P1),** to the midpoint **(P2),** to the lower endpoint **(P3).**

1. Select the Line command.
2. Draw two vertical lines as shown below.
3. Turn off "Increment Snap". (Use the Snap button or F9)
4. Select the Line command again.
5. Select the "Endpoint" object snap option using one of the methods listed on the previous page.
6. Place the cursor close to the upper endpoint of the left hand line **(P1).** *(Notice that a square appears at the end of the line. An "endpoint" tool tip should appear and the cursor snaps to the endpoint like a magnet. This is what "object snap" is all about. You are snapping the cursor to a previously drawn object.)*
7. Press the left mouse button to attach the new line to the endpoint of the previously drawn line. (Do not end the Line command yet.)
8. Now select the "Midpoint" object snap option.
9. Move the cursor to approximately the middle of the right hand vertical line **(P2).** A triangle and a "midpoint" tool tip appear, and the cursor should snap to the middle of the line like a magnet.
10. Press the left mouse button to attach the new line to the midpoint of the previously drawn line. (Do not end the Line command yet.)
11. Select the "endpoint" object snap option.
12. Move the cursor close to the lower endpoint of the left hand vertical line **(P3).**
13. Press the left mouse button to attach the new line to the endpoint of the previously drawn line.
14. Disconnect by pressing <enter>.

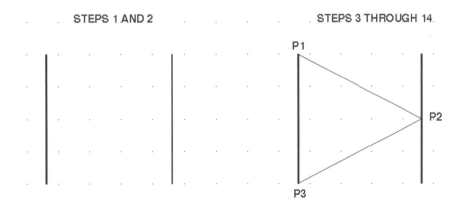

STEPS 1 AND 2   STEPS 3 THROUGH 14.

RUNNING OBJECT SNAP

RUNNING OBJECT SNAP is a method of **presetting** the **object snap options** so specific options, such as center, endpoint or midpoint, stay **active** until you **de-activate** them. When Running Object Snap is active, markers are displayed automatically as you move the cursor near the object and the cursor is drawn, to the object snap location, like a **magnet**.

For example, if you need to snap to the endpoint of 10 lines, you could preset the running object snap **endpoint** option. Then when you place the cursor near any one of the lines, a marker will appear at the endpoint and the cursor will automatically snap to the endpoint of the line. You then can move on to the next and the next and the next. Thus eliminating the necessity of invoking the object snap menu for each endpoint.

Running Object Snap can be toggled **ON** or **OFF** using the **F3** key or **clicking** on the **OSNAP** button on the status bar.

Setting Running Object Snap

1. Select the **Running Object Snap** option using one of the following:

    TYPE = OS <enter>
    PULL DOWN = TOOLS / DRAFTING SETTINGS
    Right Click on the OSNAP tile, on the Status Bar, and select SETTINGS.

    (The dialog box below will appear.)

2. Select the **OBJECT SNAP** tab.

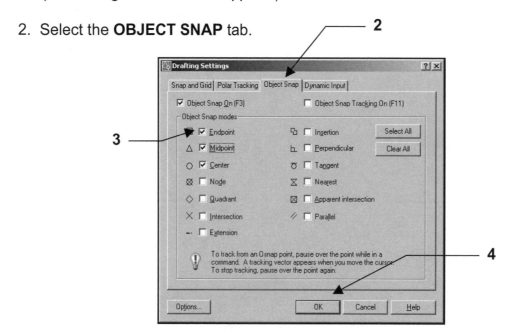

3. Select the Object Snap desired
4. Select **OK**.

Note: Do not preset more than 3 object snaps, you will lose control of the cursor.

DRAWING SET UP

When drawing with a computer, you must "set up your drawing area" just as you would on your drawing board if you were drawing with pencil and paper. You must decide what size your paper will be, what type of scale you will use (feet and inches or decimals, etc) and how precise you need to be. In CAD these decisions are called "Setting the **Drawing Limits, Units** and **Precision**".

DRAWING LIMITS
Consider the drawing limits as the size of the paper you will be drawing on. You will first be asked to define where the lower left corner should be placed, then the upper right corner, similar to drawing a Rectangle. An 11 x 17 piece of paper would have a **lower left corner** of 0,0 and an **upper right corner** of 17, 11. *(17 is the horizontal measurement or X-axis and 11 is the vertical measurement or Y-axis.)*

HOW TO SET THE DRAWING LIMITS
1. Select the **DRAWING LIMITS** command using one of the following:

    TYPE = LIMITS <enter>
    PULLDOWN = FORMAT / DRAWING LIMITS
    TOOLBARS = NONE

2. The following will appear on the command line:

    Command: '_limits
    Reset Model space limits:
    Specify lower left corner or [ON/OFF] <0.000,0.000>:

3. Type the X,Y coordinates **0, 0** for the lower left corner location of your piece of paper then press <enter>.

4. The command line will now read:

    Specify upper right corner <12.000,9.000>:

5. Type the X,Y coordinates **17, 11** for the upper right corner of your piece of paper then press <enter>.

6. **This next step is very important:** Select **VIEW / ZOOM / ALL** to make the screen display the new drawing limits.

A personal note: *I love drawing limits. I can remember when I used to draw with pencil and paper and I would find that I had under estimated the size of the paper needed. With drawing limits you merely type a new size and magically the paper gets bigger or smaller. So now you can look forward to beautiful drawings with unlimited space. We will learn how to print any size drawing later in this workbook. Be patient for now.*

---

continued next page....

DRAWING SET UP continued

UNITS AND PRECISION
You now need to select what unit of measurement you want to work with.
Such as: Decimal (0.000) or Architectural (0'-0").
Next you should select how precise you want the measurements. This means, do you
want the measurement rounded off to a 3 place decimal or the nearest 1/8".

HOW TO SET THE UNITS AND PRECISION.
1. Select the **UNITS** command using one of the following:

> **TYPE = UNITS <enter>**
> **PULLDOWN = FORMAT / UNITS**
> **TOOLBAR = NONE**

(The dialog box below will appear.)

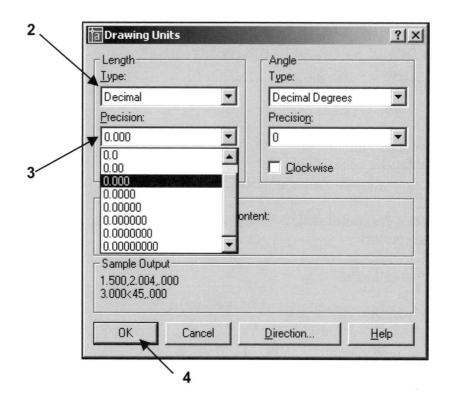

2. Select the appropriate **TYPE** such as: decimals or architectural.

3. Select the appropriate **PRECISION** associated with the "type".

4. Select the **OK** button to save your selections.

Easy, yes?

ZOOM

The **ZOOM** command is used to move closer or farther away to an object.

The following is an example of **Zoom / Window** to zoom in closer to an object.

1. Select the Zoom command by using one of the following:

> **TYPING = Z <enter>**
> **PULLDOWN = VIEW / ZOOM**
> **TOOLBAR = STANDARD**

2. Select the **"Window"** option and draw a window around the area you wish to magnify by moving the cursor to the lower left area of the object(s) and left click. Then move the cursor diagonally to form a square shape around the objects, and left click again. *(Do not hold the left mouse button down while moving the cursor, just click it at the first and diagonal corners of the square shape)*

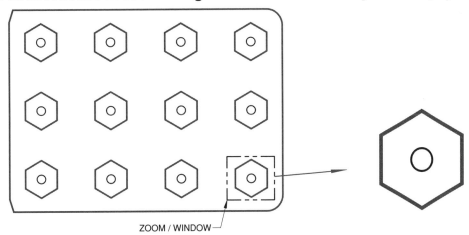

ZOOM / WINDOW

Additional Zoom options described below. (Try them)

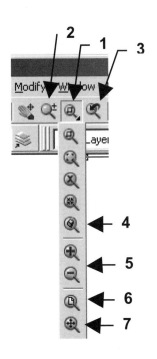

1. **WINDOW** = zoom in on an area by specifying a window (rectangle) around the area.

2. **REAL TIME** = Interactive Zoom. You can zoom in or out by moving the cursor vertically up or down while pressing the left mouse button. To stop, press the Esc key.

3. **PREVIOUS** = returns the screen to the previous display. (Limited to 10 previous displays)

4. **OBJECT** = zooms in on a selected object

5. **IN or OUT** = moves in 2X or out 2X

6. **All** = Changes the screen to the size of the drawing limits. If you have objects outside of the drawing limits, Zoom/All will display them too.

7. **EXTENTS** = Displays all objects in the drawing file, using the smallest window possible.

EXERCISE 4A

INSTRUCTIONS:

1. Start a **New** file and select **1workbook helper.dwt**.

2. Using **FORMAT / UNITS:**
   set the units to **FRACTIONAL**
   set the precision to **1/2"**

3. Using **FORMAT / DRAWING LIMITS** set the drawing limits to:
   Lower left corner = **0,0**
   Upper right corner = **20, 15**

4. Use **VIEW / ZOOM / ALL** to make the screen adjust to the new limits

5. Turn **OFF** the **GRIDS** (F7) **SNAP** (F9) and **ORTHO** (F8).
   (Your screen should be blank and your crosshair should move freely)

6. Draw the objects below using:
   DRAW / CIRCLE (CENTER, RADIUS) and **LINE**
   OBJECT SNAP = CENTER and TANGENT

7. **Save** this drawing as: **EX4A**

Very Important: Use the Tangent option at <u>each end</u> of the line. AutoCAD needs to be told that you want <u>each end</u> of the line to be tangent to a circle.

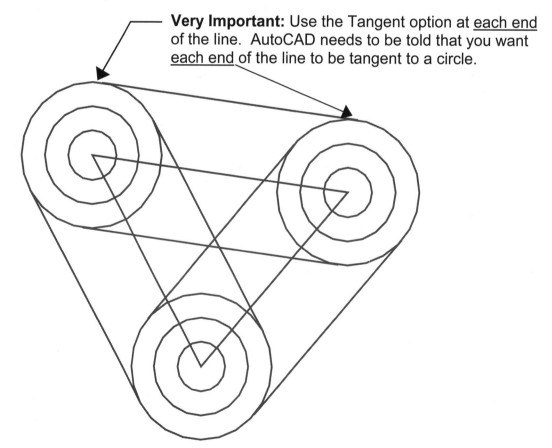

EXERCISE 4B

INSTRUCTIONS:

1. Start a **New** file and select **1workbook helper.dwt**.

2. Using **FORMAT / UNITS:**
   Set the units to **FRACTIONAL**
   Set the precision to **1/4"**

3. Using **FORMAT / DRAWING LIMITS** set the drawing limits to:
   Lower left corner = **0, 0**
   Upper right corner = **12, 9**

4. Use **VIEW / ZOOM / ALL** to make the screen adjust to the new limits.

5. Turn **OFF** the **GRIDS** (F7) **SNAP** (F9) and **ORTHO** (F8)
   (Your screen should be blank and your crosshair should move freely)

6. Draw the objects below using:
   DRAW / CIRCLE (CENTER, RADIUS) and **LINE**
   OBJECT SNAP = QUADRANT

7. **Save** this drawing as: **EX4B**

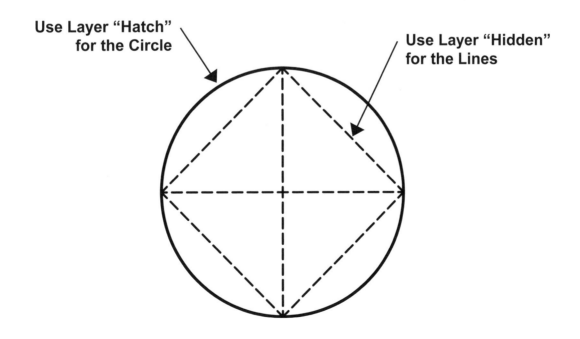

Use Layer "Hatch" for the Circle

Use Layer "Hidden" for the Lines

EXERCISE 4C

INSTRUCTIONS:

1. Start a **New** file and select **1workbook helper.dwt**.

2. Using **FORMAT / UNITS:**
    Set the units to **ARCHITECTURAL**
    Set the precision to **1/2"**

Note: A warning may appear asking you if you "are sure you want to change the units"?   Select the OK button.

3. Using **FORMAT / DRAWING LIMITS** set the drawing limits to:
    Lower left corner = **0, 0**
    Upper right corner = **25, 20**

4. Use **VIEW / ZOOM / ALL** to make the screen adjust to the new limits.

5. Turn **OFF** the **GRIDS** (F7) **SNAP** (F9) and **ORTHO** (F8)
    (Your screen should be blank and your crosshair should move freely)

6. Draw the objects below using:
    DRAW / LINE
    OBJECT SNAP = PERPENDICULAR

7. **Save** this drawing as: **EX4C**

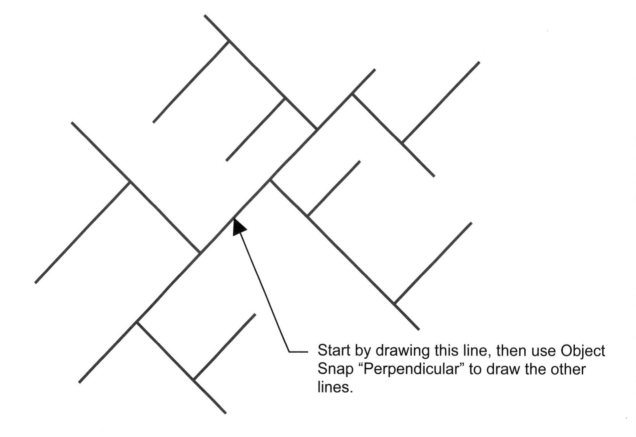

Start by drawing this line, then use Object Snap "Perpendicular" to draw the other lines.

EXERCISE 4D

INSTRUCTIONS:

1. Start a **New** file and select **1workbook helper.dwt**.

2. Using **FORMAT / UNITS:**
    Set the units to **DECIMALS**
    Set the precision to **0.00**

3. Using **FORMAT / DRAWING LIMITS** set the drawing limits to:
    Lower left corner = **0,0**
    Upper right corner = **12, 9**

4. Use **VIEW / ZOOM / ALL** to make the screen adjust to the new limits.

5. Turn **OFF** the **GRIDS** (F7) **SNAP** (F9) and **ORTHO** (F8)
    (Your screen should be blank and your crosshair should move freely)

6. Draw the Lines below using:
    DRAW / LINE
    OBJECT SNAP = MIDPOINT

7. **Save** this drawing as: **EX4D**

Start here

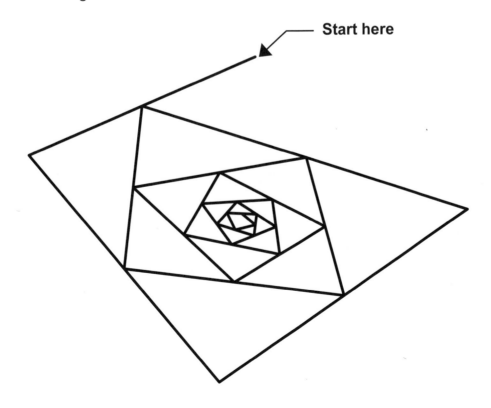

EXERCISE 4E

INSTRUCTIONS:

1. Start a **New** file and select **1workbook helper.dwt**

2. Draw the objects below using:
    DRAW / LINE
    ORTHO **ON** for Horizontal Lines
    OBJECT SNAP = ENDPOINT

3. **Save** this drawing as: **EX4E**

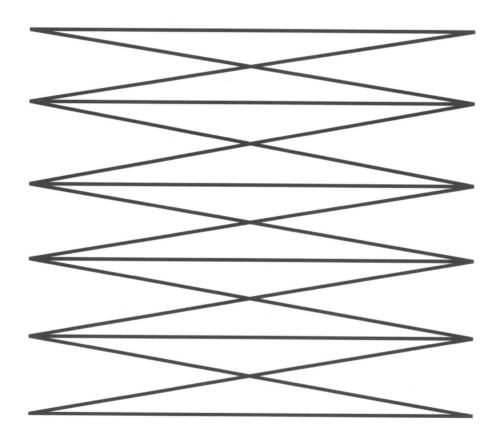

EXERCISE 4F

INSTRUCTIONS:

1. Start a **New** file and select **1workbook helper.dwt**

2. Draw the 2 vertical and 4 horizontal lines using:
   DRAW / LINE
   ORTHO (F8) = **ON**
   SNAP (F9) = **OFF**

3. Then draw the diagonal lines using:
   DRAW / LINE
   ORTHO & SNAP= **OFF**
   OBJECT SNAP = INTERSECTION

4. **Save** this drawing as: **EX4F**

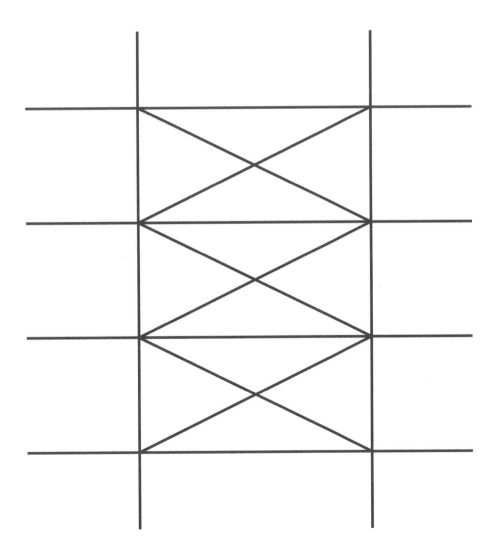

EXERCISE 4G

INSTRUCTIONS:

1. Start a **New** file and select **1workbook helper.dwt**.

2. Draw the 4 circles with the following Radii: 1, 2, 3, & 5
   (Use Object snap "Center" so all Circles have the same center)

3. Draw the LINES using:
   DRAW / LINE
   ORTHO and SNAP = **OFF**
   OBJECT SNAP = QUADRANT and TANGENT

4. Use Layers: Object and Center

5. **Save** this drawing as **EX4G**

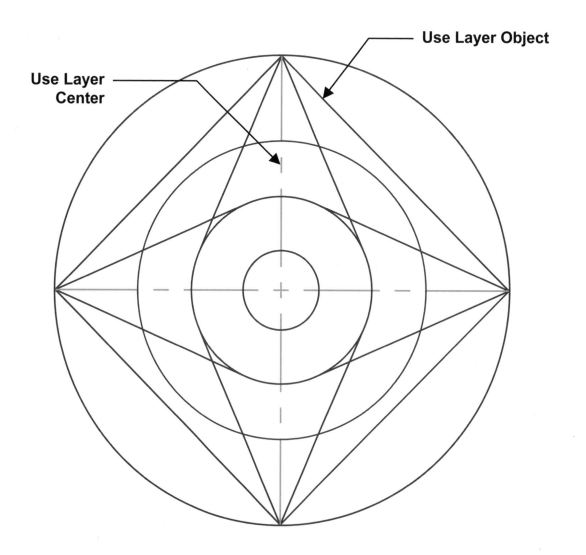

Use Layer Object

Use Layer Center

LEARNING OBJECTIVES

After completing this lesson, you will be able to:

1. Draw an Inscribed and Circumscribed Polygon.
2. Create an Ellipse using two different methods.
3. Draw an object called Donut.
4. Define a Point location.
5. Select various Point Styles.
6. Use 3 new Object Snap modes.

LESSON 5

POLYGON

A polygon is an object with multiple sides of equal length. You may specify from 3 to 1024 sides. A polygon appears to be multiple lines, but, in fact, it is one object. You can specify the edge length or the center and a radius. The radius can be drawn inside (Inscribed) an imaginary circle or outside (Circumscribed) an imaginary circle.

CENTER / RADIUS METHOD

1. Select the **Polygon** command using one of the following:

> **TYPING = POL <enter>**
> **PULLDOWN = DRAW / POLYGON**
> **TOOLBAR = DRAW**

2. The following prompts will appear on the command line:

> _polygon Enter number of sides <4>: *type number of sides <enter>*
> Specify center of polygon or [Edge]: *specify the center location*
> Enter an option [Inscribed in circle/Circumscribed about circle]<I>:*type I or C<enter>*
> Specify radius of circle: *type radius or locate with cursor.*

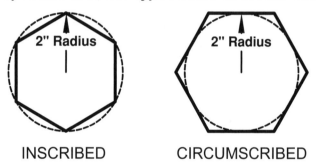

INSCRIBED   CIRCUMSCRIBED

EDGE METHOD

1. Select the **Polygon** command using one of the following:

> **TYPING = POL**
> **PULLDOWN = DRAW / POLYGON**
> **TOOLBAR = DRAW**

2. The following prompts will appear on the command line:

> _polygon Enter number of sides <4>: *type number of sides <enter>*
> Specify center of polygon or [Edge]: *type E <enter>*
> Specify first endpoint of edge: *place first endpoint of edge (P1)*
> Specify second endpoint of edge: *place second endpoint of edge (P2)*

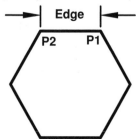

ELLIPSE

An Ellipse may be drawn by specifying the 3 points of the axes or by defining the center point and major / minor axis points.

AXIS END METHOD

1. Select the **ELLIPSE** command using one of the following:

> **TYPING =EL <enter>**
> **PULLDOWN = DRAW / ELLIPSE**
> **TOOLBAR = DRAW**

2. The following prompts will appear on the command line:

Command: _ellipse
Specify axis endpoint of ellipse or [Arc/Center]: ***place the first point of either the major or minor axis (P1).***
Specify other endpoint of axis: ***place the other point of the first axis (P2)***
Specify distance to other axis or [Rotation]: ***place the point perpendicular to the first axis (P3).***

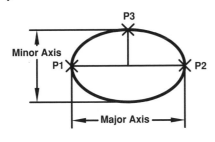 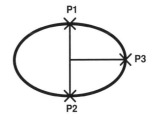

CENTER METHOD

1. Select the **ELLIPSE** command using one of the following:

> **TYPING =EL**
> **PULLDOWN = DRAW / ELLIPSE**
> **TOOLBAR = DRAW**

2. The following prompts will appear on the command line:

Command: _ellipse
Specify axis endpoint of ellipse or [Arc/Center]: ***type C <enter>***
Specify center of ellipse: ***place center of ellipse (P1)***
Specify endpoint of axis: ***place first axis endpoint (either axis) (P2)***
Specify distance to other axis or [Rotation]: ***place the endpoint perpendicular to the first axis (P3)***

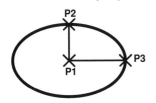

DONUT

A Donut is a circle with *width*. You will define the **Inside** and **Outside** diameters.

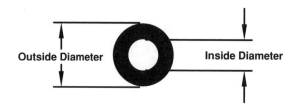

1. Select the **DONUT** command using one of the following:

   TYPING =DO<enter>
   PULLDOWN = DRAW / DONUT
   TOOLBAR = DRAW

2. The following prompts will appear on the command line:

   Command: _donut
   Specify inside diameter of donut: *type the inside diameter*
   Specify outside diameter of donut: *type the outside diameter*
   Specify center of donut or <exit>: *place the center of the first donut*
   Specify center of donut or <exit>: *place the center of the second donut or*
   <enter> to stop

   Controlling the "FILL MODE"

   1. Command: *type FILL <enter>*

   2. Enter mode [ON / OFF] <OFF>: *type ON or OFF <enter>*

FILL = ON

FILL = OFF

3. Select **VIEW / REGEN** or type *REGEN <enter>* at the command line to update the drawing to the latest changes to the *FILL* mode.

POINT

A Point is an object that has no dimension and only has location. A Point may be represented by one of many Point styles shown below. Points are basically used to locate a point of reference. A location.

You may select the "Single" or "Multiple" point option. The Single option creates one point. The Multiple option will continue until you press the ESC key.

The only Object Snap mode that can be used with Point is "**Node**". (See page 5-6)

1. Select the **POINT** command using one of the following:

>   **TYPING = PO <enter>**
>   **PULLDOWN = DRAW / POINT (Single point or Multiple point)**
>   **TOOLBAR = DRAW**

2. The following prompts will appear on the command line:

    Command: _point
    Current point modes: PDMODE=3 PDSIZE=0.000
    Specify a point: *place the point location*
    (If you selected "multiple point" you must press the "ESC" key to stop.)

TO SELECT A "POINT STYLE"

1. Select one of the following:
>   **TYPE = DDPTYPE**
>   **PULLDOWN = FORMAT / POINT STYLE**
>   **TOOLBAR = NONE**

2. This dialog box will appear.

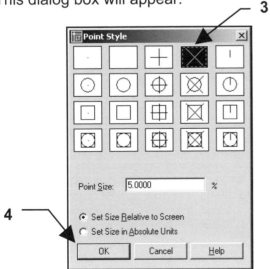

3. Select a tile.
4. Select the OK button.

MORE OBJECT SNAP

OBJECT SNAP OPTIONS:

 NODe   This option snaps to a POINT object.
Place the pick box on the POINT object.

 NEArest   Snaps to the nearest location on the nearest object.
Select the object anywhere on the object.

Mid Between 2 Points (Sorry no tool on Object Snap toolbar)

Locates a midpoint between two points

Snap to the first location and then the second. The midpoint between these 2 points will be located automatically.

EXERCISE 5A

INSTRUCTIONS:

1. Start a **New** file and select **1workbook helper.dwt**

2. Draw the Circle first, Polygon second and Lines last.

3. Draw the objects below using:
   DRAW / CIRCLE (Center, Rad)
   DRAW / POLYGON (Circumscribed)
   Ortho (F8) = **ON**
   Snap (F9) = **OFF**
   Object Snap = Center, Midpoint, Endpoint and Quadrant

4. **Save** this drawing as: **EX5A**

Draw the Circle first then place the center of the Polygon in the center of the Circle using object snap "Center"

Snap line to Midpoint of Polygon segment.

Snap Polygon to the Circle using Quadrant

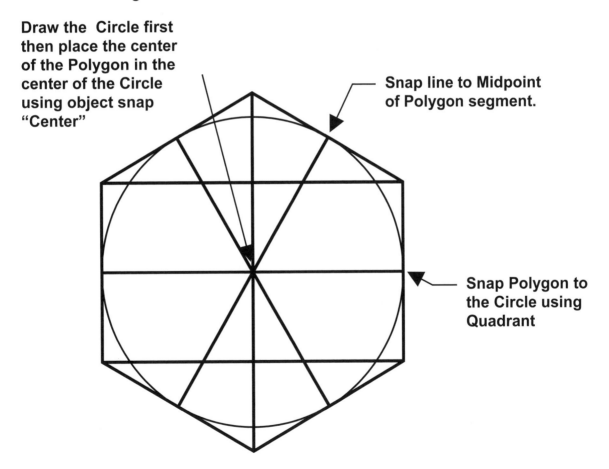

EXERCISE 5B

INSTRUCTIONS:

1. Start a **New** file and select **1workbook helper.dwt**.

2. Select the "Point Style".

3. Draw the Point first, then the Polygons.

4. Draw the objects below using:
   DRAW / POINT and **POLYGON** (Inscribed)
   Ortho and Snap **ON**
   Object Snap = Node

5. Locate the center of the Polygon by snapping to the Point using Object snap: **Node**

6. **Save** this drawing as: **EX5B**

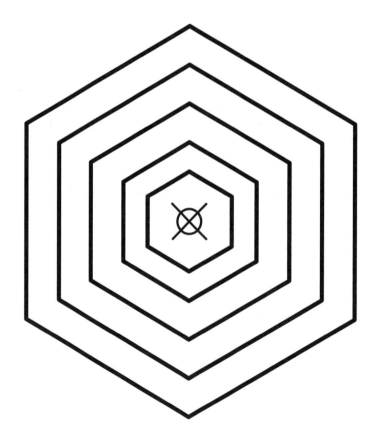

Note: Your "Point" may not appear as large as the one above.

EXERCISE 5C

INSTRUCTIONS:

1. Start a **New** file and select **1workbook helper.dwt**.
2. Draw the objects below using:
   DRAW / ELLIPSE (Axis, End) **LINES** and **DONUT**
   Ortho **ON** and Snap **OFF**
   Object Snap = Perpendicular, Quadrant and Center
3. Have fun with this one.
4. **Save** this drawing as: **EX5C**.

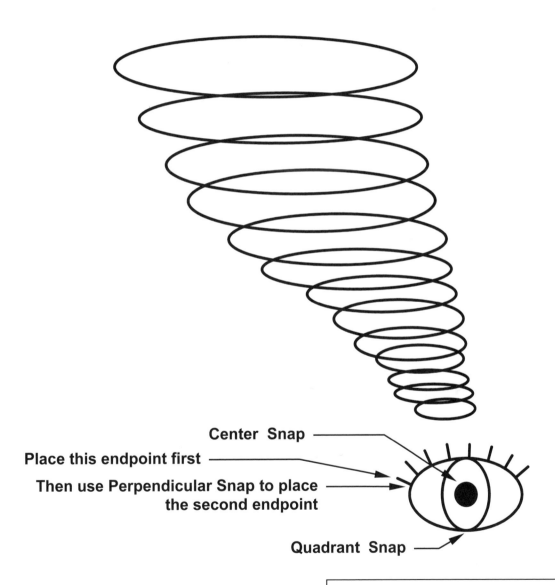

Center Snap

Place this endpoint first

Then use Perpendicular Snap to place
the second endpoint

Quadrant Snap

Note: This is the "eye" of the TORNADO.
(Just a little fun!)

EXERCISE 5D

INSTRUCTIONS:

1. Start a **New** file and select **1workbook helper.dwt**.
2. Select the Point Style.
3. Draw the Point first, then the Ellipse
4. Draw the objects below using:
    DRAW / POINT
    DRAW / ELLIPSE (Use "Center" method)
    Ortho **ON** and Snap **OFF**
    Object Snap = Quadrant and Node
4. **Save** this drawing as: **EX5D**

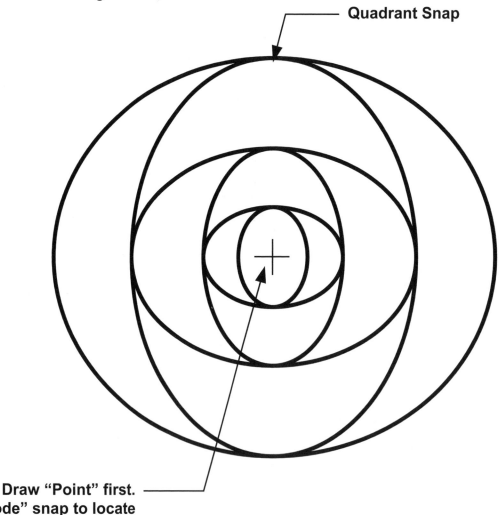

Quadrant Snap

Draw "Point" first.
Use "Node" snap to locate
the center of the ellipse.

EXERCISE 5E

INSTRUCTIONS:

STEP 1.
1. Start a **New** file and select **1workbook helper.dwt**.
2. Draw the objects below using:
   DRAW / DONUT
   Ortho and Snap **OFF**
   Object Snap = Center

3. **Save** this drawing as **EX5E**

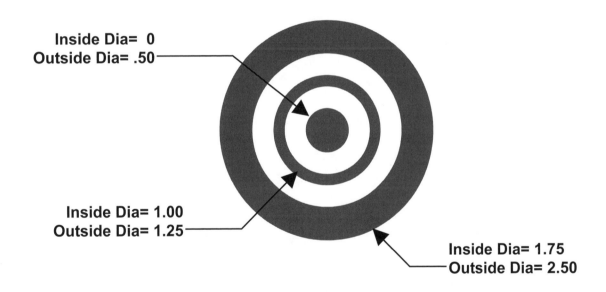

Inside Dia= 0
Outside Dia= .50

Inside Dia= 1.00
Outside Dia= 1.25

Inside Dia= 1.75
Outside Dia= 2.50

EXERCISE 5F

INSTRUCTIONS:

1. Start a **New** file and select **1workbook helper.dwt**.
2. Draw the objects below using:
   DRAW / POINT, POLYGON, ELLIPSE (center) and **DONUT**
   Ortho **ON** and Snap **OFF**
   Object Snap = Node, Endpoint and Midpoint
3. **Save** this drawing as: **EX5F**

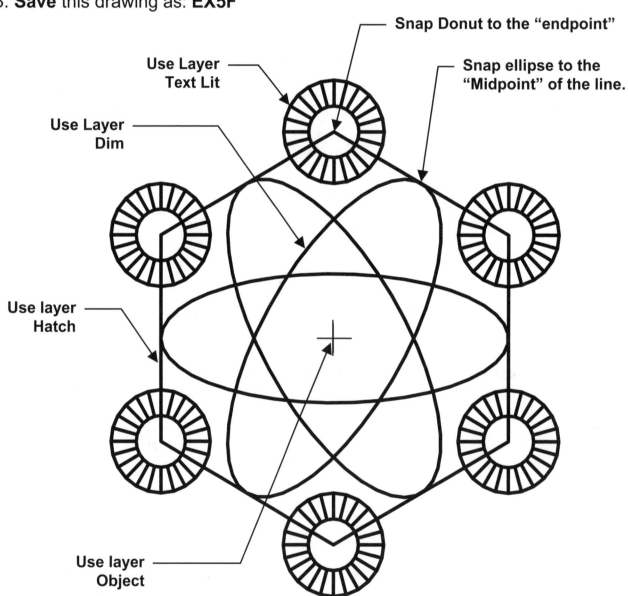

Snap Donut to the "endpoint"

Snap ellipse to the "Midpoint" of the line.

Use Layer Text Lit

Use Layer Dim

Use layer Hatch

Use layer Object

EXERCISE 5G

INSTRUCTIONS:

1. Start a **New** file and select **1workbook helper.dwt**.

2. Draw the zig zag line approximately as shown.

3. Set the "running object snap" setting to "Endpoint".

4. Draw the circles at the "Midpoint between 2 points on the zig zag line.

   Use the "Mid Between 2 Points" method described on page 5-6.
   a. Select the Circle command.
   b. Type MTP <enter>
   c. Snap to the endpoints (P1 and P2)
   (The center of the circle should be exactly midway between the 2 points)
   d. Place the Radius of the circle.

5. **Save** this drawing as: **EX5G**

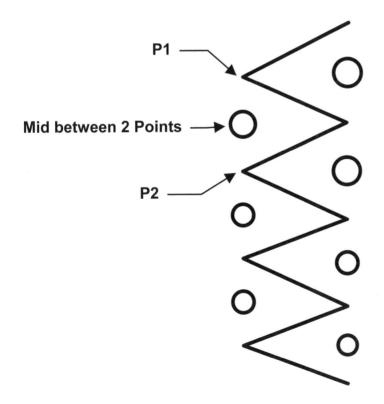

NOTES:

LEARNING OBJECTIVES

After completing this lesson, you will be able to:

1. Use the 4 Break command options.
2. Trim an object to a cutting edge.
3. Extend an object to a boundary.
4. Move an object(s) to a new location.
5. Explode objects into their primitive entities.

LESSON 6

BREAK

The *BREAK* command allows you to break a space in an object, break the end off an object or split a line in two. I think of it as taking a bite out of an object.
There are 4 break methods, as described below.

You may select the **BREAK** command by using one of the following:

TYPING = BR
PULLDOWN = MODIFY / BREAK
TOOLBAR = MODIFY

METHOD 1
How to break one object into two separate objects with no visible space in between. (Use this method if the location of the break is not important)

a. Draw a line.
b. Select the **BREAK** command using one of the methods listed above.
c. _break Select object: *pick the break location (P1) by clicking on it.*
d. Specify second break point or [First point]: *type @ <enter>.*
   (This will duplicate the last point.)
e. Now if you click on one end of the line you will see that there are 2 lines instead of just one.

METHOD 2
This method is the same as method 1, however, <u>use this method if the location of the break is very specific.</u>

a. Select the **BREAK** command using one of the methods listed above.
b. _break Select objects: *select the object to break (P1).*
c. Specify second break point or [First point]: *type F <enter>.*
d. Specify first break point: *select break location (P2) accurately.*
e. Specify second break point: *type @ <enter>.*

Method 2 can also be accomplished easily by selecting the "Break at point" icon.   But I wanted you to understand how it works.

BREAK (continued)

METHOD 3
Take a bite out of an object. (Use this method if the location of the BREAK is not important.)
(This is the Default option.)

a. Select the **BREAK** command.
b. _break Select object: *pick the first break location (P1).*
c. Specify second break point or [First point]: *pick the second break location (P2).*

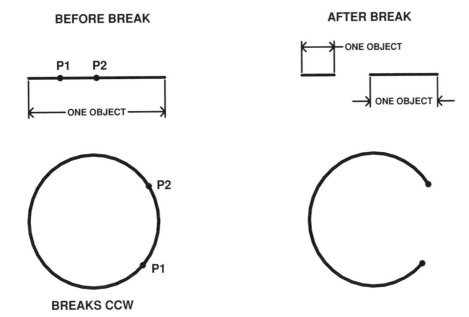

Note: Circles can't be broken with "1 point". You must use 2 points.

METHOD 4
This method is the same as method 3, however, <u>use this method if the location of the break is very specific.</u>

a. Select the **BREAK** command.
b. _break Select objects: *select the object to break (P1) anywhere on the object.*
c. Specify second break point or [First point]: *type F <enter>.*
d. Specify first break point: *select the first break location (P2) accurately.*
e. Specify second break point: *select the second break location (P3) accurately.*

TRIM

The **TRIM** command is used to trim an object to a **cutting edge**. You first select the "Cutting Edge" and then select the part of the object you want to trim. The object to be trimmed must actually intersect the cutting edge or could intersect if the objects were infinite in length.

1. Select the Trim command using one of the following:

> **TYPE = TR**
> **PULLDOWN = MODIFY / TRIM**
> **TOOLBAR = MODIFY**

2. The following will appear on the command line:

Command: _trim
Current settings: Projection = UCS   Edge = Extend
Select cutting edges ...
Select objects or <select all>: ***select cutting edge(s) by clicking on the object (P1 below)***
Select objects: ***stop selecting cutting edges by pressing the <enter> key***
Select object to trim or shift-select to extend or
[Fence/Crossing/Project/Edge/eRase/Undo]: ***select the object that you want to trim. (P2)***
(Select the part of the object that you want to disappear, not the part you want to remain)
Select object to trim or [Fence/Crossing/Project/Edge/eRase/Undo]: ***press <enter> to stop***

> Note: You may toggle between Trim and Extend (page 6-5). Hold down the shift key and the Extend command is activated. Release the shift key and you return to Trim.

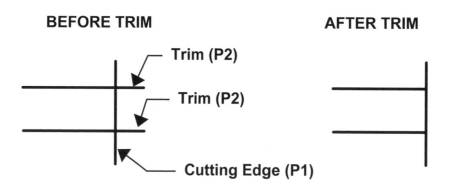

BEFORE TRIM   **AFTER TRIM**

— Trim (P2)
— Trim (P2)
— Cutting Edge (P1)

Fence See page 6-14
Edge See page 6-5
Project See page 6-5

Crossing You may select objects using a Crossing Window.

eRase You may erase an object instead of trimming while in the Trim command.

Undo You may "undo" the last trimmed object while in the Trim command.

EXTEND

The **EXTEND** command is used to extend an object to a **boundary**. The object to be extended must actually or theoretically intersect the boundary.

1. Select the **EXTEND** command using one of the following:

   TYPE = EX
   PULLDOWN = MODIFY / EXTEND
   TOOLBAR = MODIFY

2. The following will appear on the command line:

Command: _extend
Current settings: Projection = UCS   Edge = Extend
Select boundary edges ...
Select objects or <select all>: *select boundary (P1below) by clicking on the object.*
Select objects: *stop selecting boundaries by selecting <enter>.*
Select object to extend or shift-select to Trim or
[Fence/Crossing/Project/Edge/Undo]:*select the object that you want to extend (P2 and P3). (Select the end of the object that you want to extend.)*
Select object to extend or [Fence/Crossing/Project/Edge/Undo]:*stop selecting objects*
                                                    by selecting <enter>.

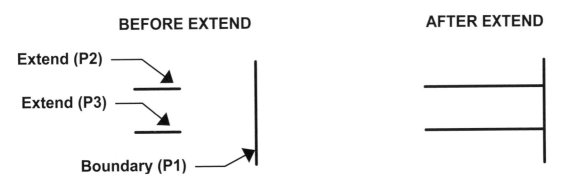

BEFORE EXTEND   **AFTER EXTEND**

Extend (P2)
Extend (P3)
Boundary (P1)

> You may toggle between Extend and Trim (page 6-4). Hold down the shift key and the Trim command is activated. Release the shift key and you return to Extend.

Fence See page 6-14
Crossing See page 6-4
Project Same as Edge except used only in "3D".
Edge (Extend or No Extend)
In the **"Extend"** mode, (default mode) the boundary and the Objects to be extended need only imaginarily intersect if the objects were infinite in length.
In the **"No Extend"** mode the boundary and the objects to be extended must visibly intersect.
Undo See page 6-4

MOVE

The MOVE command is used to move object(s) from their current location (basepoint) to a new location (second displacement point).

1. Select the Move command using one of the following:

    TYPE = M
    PULLDOWN = MODIFY / MOVE
    TOOLBAR = MODIFY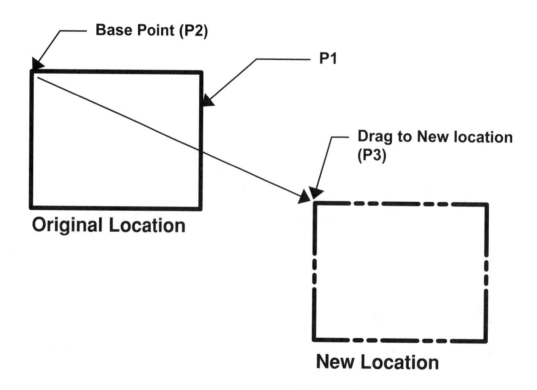

2. The following will appear on the command line:

    Command: _move
    Select objects: *select the object(s) you want to move (P1).*
    Select objects: *stop selecting object(s) by selecting <enter>.*
    Specify base point or displacement: *select a location (P2) (usually on the object).*
    Specify second point of displacement or <use first point as displacement>: *move the object to it's new location (P3) and left click.*

Note: If you press <enter> instead of actually picking a new location (P3), Autocad will send it into <u>Outer Space</u>. If this happens, press U <enter> or select the " undo" icon *and try again.*

Base Point (P2)

P1

Drag to New location (P3)

Original Location

New Location

EXPLODE

The EXPLODE command changes (explodes) an object into its primitive objects. For example: A rectangle is originally one object. If you explode it, it changes into 4 lines. Visually you will not be able to see the change unless you select one of the lines.

1. Select the Explode command by using one of the following:

   TYPE =X
   PULLDOWN = MODIFY / EXPLODE
   TOOLBAR = MODIFY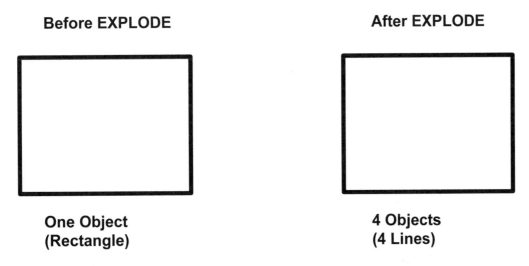

2. The following will appear on the command line:

Command: _explode
Select objects: *select the object(s) you want to explode.*
Select objects: *select <enter>.*

Before EXPLODE   **After EXPLODE**

One Object   **4 Objects**
(Rectangle)   **(4 Lines)**

(Notice there is no visible difference. But now you have 4 lines instead of 1 Rectangle)

Try this:
Draw a rectangle and then click on it. The entire object highlights.
Now explode the rectangle, then click on it again. Only the line you clicked on should be highlighted. Each line that forms the rectangular shape is now an individual object.

EXERCISE 6A

INSTRUCTIONS:

1. Start a **New** file and select **1workbook helper.dwt**.
2. Draw the objects below:
   Locate the center of the circles on the rectangle line
   using object snap = **Nearest and Midpoint**

   Ortho and Snap = **OFF**

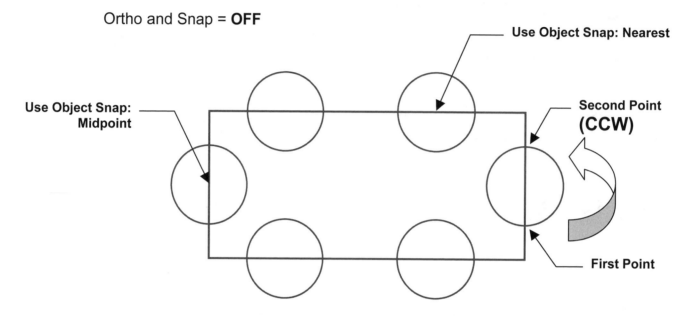

Use Object Snap: Nearest

**Use Object Snap:
Midpoint**

**Second Point
(CCW)**

First Point

3. Modify the drawing above to look like the drawing below using:
   MODIFY / BREAK (Refer to 6-3, use Method 4)
   OBJECT SNAP = INTERSECTION

4. Remember the Circles break **CCW**.

5. **Save** this drawing as **EX6A**.

Note: If your circles are not breaking where you expected, turn off "Running Object Snap" (pg. 4-4) and try again.

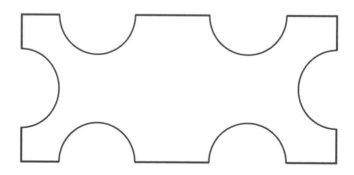

EXERCISE 6B

INSTRUCTIONS:

1. Start a **New** file and select **1workbook helper.dwt**.
2. Draw the objects below:
   OBJECT SNAP = CENTER
   ORTHO and SNAP **OFF**

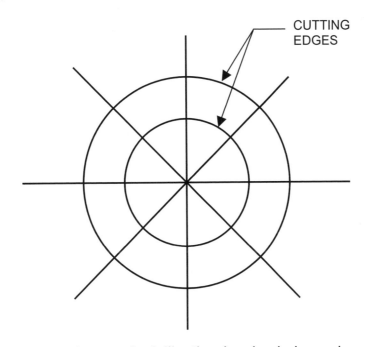

CUTTING
EDGES

3. Modify the drawing above to look like the drawing below using:
   MODIFY / TRIM
4. **Save** this drawing as **EX6B.**

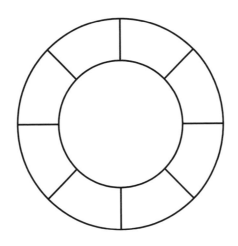

EXERCISE 6C

INSTRUCTIONS:

1. Start a **New** file and select **1workbook helper.dwt**.

2. Draw the LINES below exactly as shown.
   ORTHO and SNAP **ON**

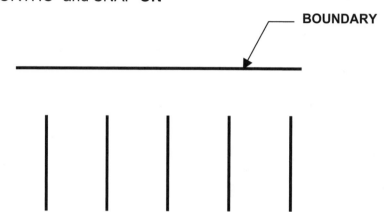

BOUNDARY

3. Modify the drawing above to look like the drawing below using:
   MODIFY / EXTEND

4. **Save** this drawing as **EX6C.**

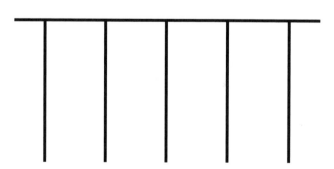

EXERCISE 6D

INSTRUCTIONS:

1. Start a **New** file and select **1workbook helper.dwt**.

2. Draw the drawing below using:
   RECTANGLE, CIRCLES and LINES (FOR THE X'S)

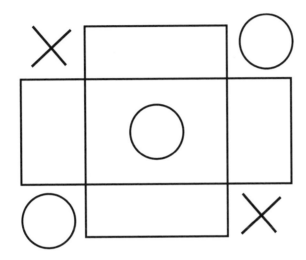

3. Modify the drawing above to look like the drawing below using:
   MODIFY / EXPLODE and ERASE

> **Note: You will not be able to erase the lines from the rectangle without Exploding the rectangle first. Now do you see why you need the Explode command?**

4. **Save** this drawing as **EX6D.**

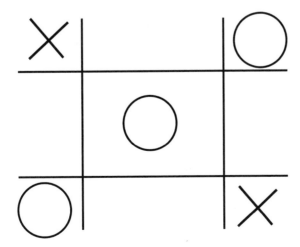

EXERCISE 6E

INSTRUCTIONS:

1. Start a **New** file and select **1workbook helper.dwt**.

2. Draw the drawing below using:
   CIRCLES, POINT and LINES.

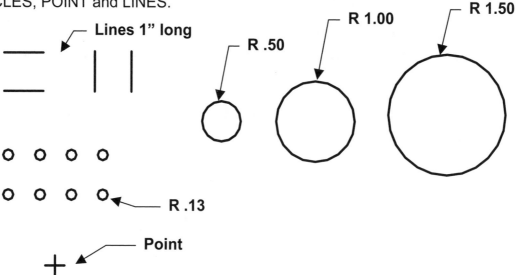

3. Assemble the objects as shown below using:
   MODIFY / MOVE
   OBJECT SNAP = CENTER, INTERSECTION, NODE, ENDPT and QUADRANT

Take your time and think about which object snap to use for each "basepoint" and "second point of displacement".

---

4. **Save** this drawing as **EX6E.**

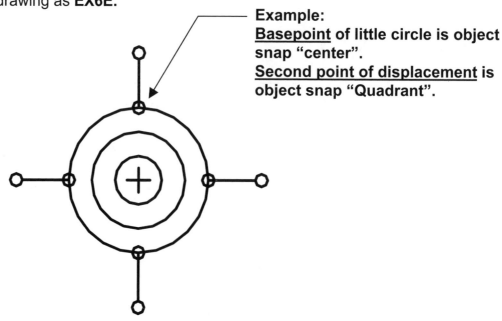

Example:
Basepoint of little circle is object snap "center".
Second point of displacement is object snap "Quadrant".

EXERCISE 6F

INSTRUCTIONS:

1. Start a **New** file and select **1workbook helper.dwt**.

2. Draw the drawing below using all the commands you have learned in the previous lessons. Use whatever layers you want and set snap and ortho on or off depending on the application.
   Be as creative as you like.

3. Save this drawing as: **EX6F**

EXERCISE 6G

INSTRUCTIONS:

1. Start a **New** file and select **1workbook helper.dwt**.

> The following exercise is exactly like exercise 6C but I want to show you a new method for selecting multiple objects without clicking on each object.

2. Draw the LINES below exactly as shown.
   ORTHO and SNAP **ON**

3. Select the "Extend" command.
4. Select the "Boundary" <enter>
5. Now instead of clicking on each vertical line, type F <enter>.
6. Place your cursor approximately at location P1 and click.
7. Place your cursor approximately at location P2 and click.
8. Press <enter> <enter>.
9. **Save** this drawing as: **EX6G.**

Note:
Be careful to place P1 and P2 above the midpoint of the vertical line. That tells AutoCAD that you want to extend the lines up. If you place P1 and P2 below the midpoint of the vertical line, AutoCAD will look for a boundary below. This will confuse AutoCAD because you did not select a boundary below.

LEARNING OBJECTIVES

After completing this lesson, you will be able to:

1. Copy objects.
2. Make a mirrored image of one or more objects.
3. Add rounded corners to rectangular objects and lines.
4. Add angles to corners.

LESSON 7

COPY

The **COPY** command creates a duplicate set of the objects selected. The COPY command is similar to the MOVE command. You must 1. select the objects to be copied, 2. select a base point and a 3. new location. The difference is, the Move command merely moves the objects to a new location. The Copy command makes a copy and you select the location for the new copy.

Select the Copy command using one of the following commands:

TYPE = CO
PULLDOWN = MODIFY / COPY
TOOLBAR = MODIFY

Command: _copy
Select objects: *select the objects you want to copy*
Select objects: *stop selecting objects by selecting <enter>*
Specify base point or displacement: *select a base point (P1) (usually on the object)*
Specify second point of displacement or <use first point as displacement>: *select the new location (P2) for the first copy*
Specify second point or [Exit / Undo] <Exit>: *select the new location (P2) for the next copy or press <enter> to exit.*

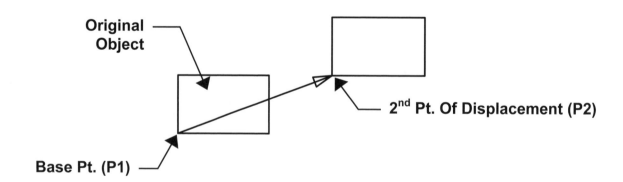

Original Object

Base Pt. (P1)

2ⁿᵈ Pt. Of Displacement (P2)

MIRROR

The MIRROR command allows you to make a mirrored image of any objects you select. You can use this command for creating right / left hand parts. You can draw a symmetrical object more efficiently by only drawing half of it.

Select the **MIRROR** command using one of the following:

TYPE = MI
PULLDOWN = MODIFY / MIRROR
TOOLBARS = MODIFY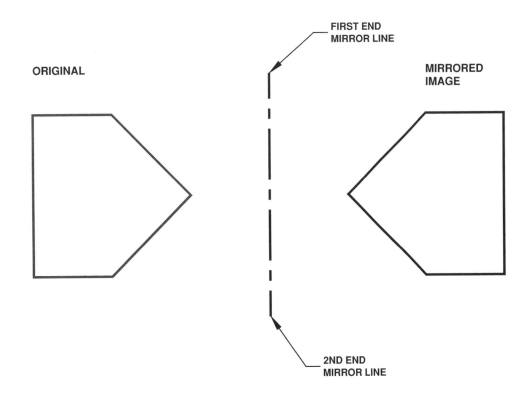

The following will appear on the command line:

Command: _mirror
Select objects: ***select the objects to be mirrored***
Select objects: ***stop selecting objects by selecting <enter>***
Specify first point of mirror line: ***select the first end of the mirror line***
Specify second point of mirror line: ***select the second end of the mirror line***
Erase source objects? [Yes/No] <N>: ***select Y or N***

ORIGINAL

FIRST END
MIRROR LINE

MIRRORED
IMAGE

2ND END
MIRROR LINE

How to control text when using the Mirror command:
1. At the command line type   **mirrtext <enter>**
2. Select the setting by typing 0 or 1 <enter>.
0 = Retains text direction   1 = Mirrors the text (default setting)

| MIRRTEXT SETTING 0 | MIRRTEXT SETTING 0 |
| MIRRTEXT SETTING 1 | MIRRTEXT SETTING 1 |

FILLET

The FILLET command will create a radius between two objects. The objects do not have to be touching. If two parallel lines are selected, it will construct a full radius.

RADIUS A CORNER

1. Select the FILLET command using one of the following:

> **TYPE = F**
> **PULLDOWN = MODIFY / FILLET**
> **TOOLBAR = MODIFY**

The following will appear on the command line:

2. SET THE RADIUS OF THE FILLET
Command: _fillet
Current settings: Mode = TRIM, Radius = 0.000
Select first object or [Undo/Polyline/Radius/Trim/Multiple]: *type "R" <enter>*
Specify fillet radius <0.000>: *type the radius <enter>*

3. NOW FILLET THE OBJECTS
Select first object or [Undo/Polyline/Radius/Trim/Multiple]: *select the first object to be filleted*
Select second object or shift-select to apply corner: *select the second object to be filleted*

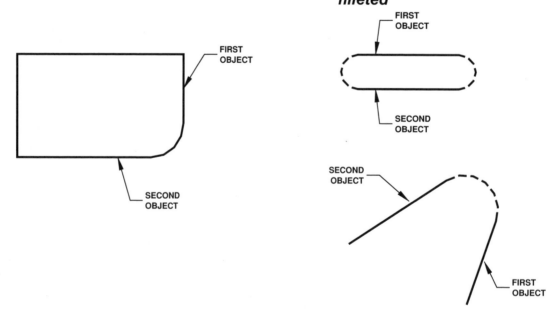

Polyline: This option allows you to fillet all intersections of a Polyline in one operation, such as all 4 corners of a rectangle.

Trim: This option controls whether the original lines are trimmed to the end of the Arc or remain the original length. (Set to Trim or No trim)

Multiple: Repeats the fillet command until you press <enter> or esc key.

FILLET continued…

The FILLET command may also be used to create a square corner.

SQUARE CORNER

1. Select the FILLET command using one of the following:

   TYPE = F
   PULLDOWN = MODIFY / FILLET
   TOOLBAR = MODIFY

The following will appear on the command line:

2. Select the two lines to form the square corner
Select first object or [Undo/Polyline/Radius/Trim/Multiple]: *select the first object (P1)*

Select second object or shift-select to apply corner: *Hold the shift key down while selecting the second object (P2)*

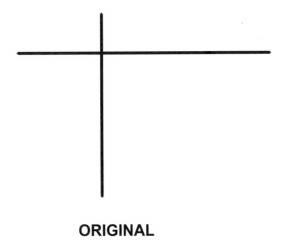

ORIGINAL

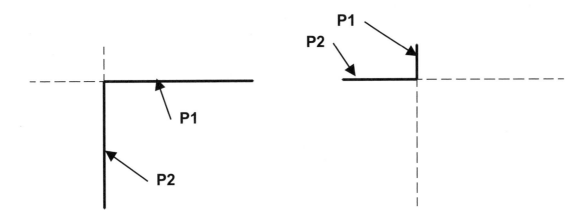

Note: The corner direction depends on which end of the object you select.

CHAMFER

The **CHAMFER** command allows you to create a chamfered corner on two lines. There are two methods: **Distance (page 7-6) and Angle (page 7-7)**.

DISTANCE METHOD
1. Select the CHAMFER command using one of the following:

> **TYPE = CHA**
> **PULLDOWN = MODIFY / CHAMFER**
> **TOOLBAR = MODIFY**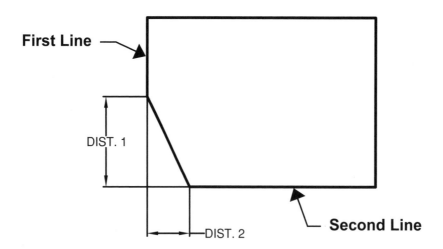

(Distance Method requires input of a distance for each side of the corner.)

Command: _chamfer
(TRIM mode) Current chamfer Dist1 = 0.000, Dist2 = 0.000
Select first line or [Undo/Polyline/Distance/Angle/Trim/mEthod/Multiple]: *select "D"<enter>.*
Specify first chamfer distance <0.000>: *type the distance <enter>.*
Specify second chamfer distance <1.000>: *type the distance <enter>.*

2. **NOW CHAMFER THE OBJECT**
Select first line or [Undo/Polyline/Distance/Angle/Trim/mEthod/Multiple]:
select the (First Line) to be chamfered (dist 1 side).
Select second line or shift-select to apply corner: *select the (Second Line) to be chamfered (dist 2 side).*

First Line
DIST. 1
DIST. 2
Second Line

Polyline: This option allows you to Chamfer all intersections of a Polyline in one operation. Such as all 4 corners of a rectangle.

Trim: This option controls whether the original lines are trimmed or remain after the corners are chamfered. (Set to Trim or No trim.)

mEthod: Allows you to switch between **Distance** and **Angle** method. The distance or angle must have been set previously.

Multiple: Repeats the Chamfer command until you press <enter> or esc key.

Note: You may use the "apply corner" also. Refer to page 7-5

CHAMFER (continued)

ANGLE METHOD

1. Select the CHAMFER command

   (Angle method requires input for the length of the line and an angle)

   Command: _chamfer
   (TRIM mode) Current chamfer Dist1 = 1.000, Dist2 = 1.000
   Select first line or [Undo/Polyline/Distance/Angle/Trim/method/Multiple]:
   select "A" <enter>
   Specify chamfer length on the first line <0.000>: **type the chamfer length <enter>**
   (dist 1)
   Specify chamfer angle from the first line <0>: **type the angle <enter>**

2. **NOW CHAMFER THE OBJECT**
   Select first line or [Undo/Polyline/Distance/Angle/Trim/mEthod/Multiple]: **select the (First
   Line) to be chamfered. (the length side)**
   Select second line or shift-select to apply corner: **select the (second line) to be
   chamfered. (the Angle side)**

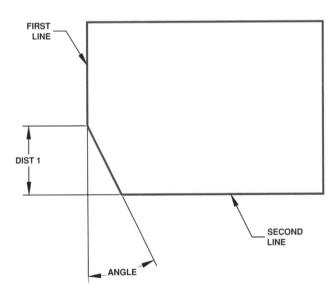

EXERCISE 7A

INSTRUCTIONS:

1. Start a **New** file using **1workbook helper.dwt**.
2. **Draw** the rectangle below using:
   Draw / Rectangle
3. **Round** the corners using the FILLET command
4. **Save** this drawing as **EX7A**

BEFORE FILLET

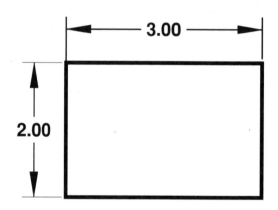

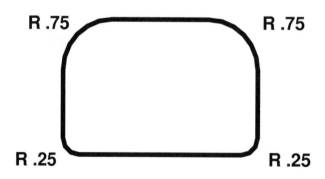

R .75   R .75

R .25   R .25

EXERCISE 7B

INSTRUCTIONS:

1. Start a **New** file **1workbook helper.dwt**.
2. **Draw** the rectangles below using:
   Draw / Rectangle
3. **CHAMFER** the corners using the CHAMFER command
4. **Save** this drawing as **EX7B**

Before Chamfer

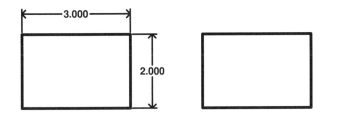 

After Chamfer

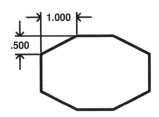 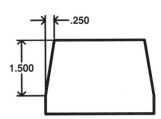 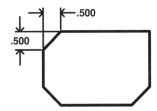

EXERCISE 7C

INSTRUCTIONS:

1. Start a **New** file and select **1workbook helper.dwt**.
2. **Draw** the rectangle below using:
   Draw / Rectangle
3. Chamfer the corners using the **CHAMFER** command
4. **Save** this drawing as **EX7C**

BEFORE CHAMFER

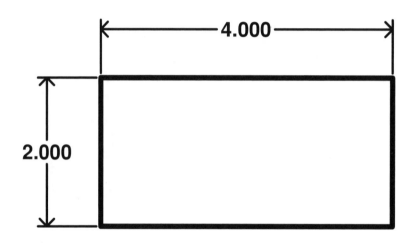

AFTER CHAMFER

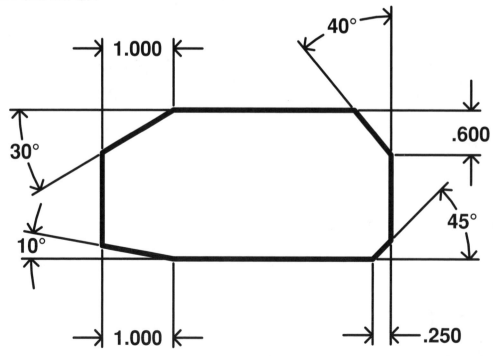

EXERCISE 7D

INSTRUCTIONS:

1. Start a **New** file and select **1workbook helper.dwt**.
2. **Draw** the Lines and one Circle as shown below

BEFORE COPY

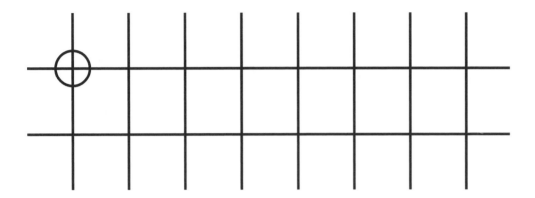

3. Complete the drawing as shown below using:
   a. Copy
   b. Select the Circle
   c. Select the basepoint on the Original circle. (Notice the basepoints are not all the same)
   d. Select the New location (2nd point of displacement)

4. Save this drawing as **EX7D**

AFTER COPY

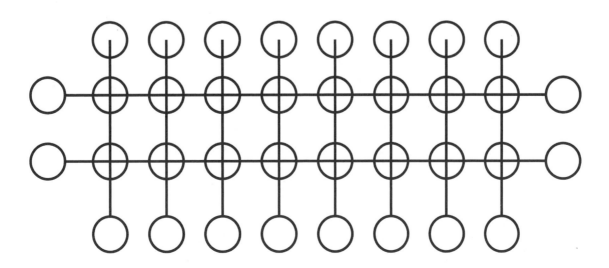

EXERCISE 7E

INSTRUCTIONS:

1. Start a **New** file and select **1workbook helper.dwt**.
2. Draw the half house below using:
   Lines, Circles and Rectangles
3. Use at least 4 different layers.

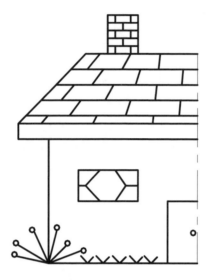

4. Now create a MIRRORED IMAGE of the half house using:
   a. Modify / Mirror
   b. Select the objects to be mirrored
   c. Select the 1st endpoint of the mirror line
   d. Select the 2nd endpoint of the mirror line
   e. Answer **No** to "Delete old objects?"

5. Save this drawing as: **EX7E**

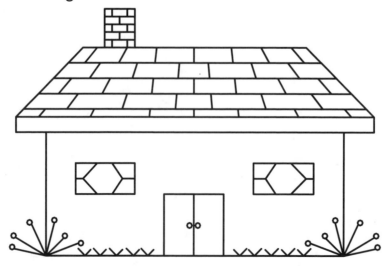

LEARNING OBJECTIVES

After completing this lesson, you will be able to:

1. Add a "Single Line" of text to your drawing.
2. Add a paragraph, using "Multiline Text".
3. Control tabs, indents and line spacing.
4. Edit text already in the drawing.
5. Mask text background
6. Scale Text

LESSON 8

SINGLE LINE TEXT

SINGLE LINE TEXT allows you to draw one or more lines of text. The text is visible as you type. To place the text in the drawing, you may use the default **START POINT** (the lower left corner of the text), or use one of the many styles of justification described on the next page.

USING THE DEFAULT START POINT

1. Select the **SINGLE LINE TEXT** command using one of the following:

TYPE = DT or TEXT
PULLDOWN = DRAW / TEXT / SINGLE LINE TEXT
TOOLBAR = DRAW AI

    Command: _dtext
    Current text style: "STANDARD" Text height: 0.250
2. Specify start point of text or [Justify/Style]: ***Place the cursor where the text should start and left click.***
3. Specify height <0.250>: ***type the height of your text***
4. Specify rotation angle of text <0>: ***type the rotation angle then <enter>***
5. Enter text: ***type the text string; press enter at the end of the sentence***
6. Enter text: ***type the text string; press enter at the end of the sentence***
7. Enter text: ***type the next sentence or press <enter> to stop***

USING JUSTIFICATION

If you need to be very specific, where your text is located, you must use the Justification option. For example if you want your text in the middle of a rectangular box, you would use the justification option "Middle".

The following is an example of Middle justification.

1. Draw a Rectangle 6" wide and 3" high.
2. Draw a Diagonal line from one corner to the diagonal corner.

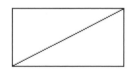

  3. Select the SINGLE LINE TEXT command
     Command: _dtext
     Current text style: "STANDARD" Text height: 0.250
  4. Specify start point of text or[Justify/Style]: ***type "J"***
  5. Enter an option [Align/Fit/Center/Middle/Right/TL/ TC/TR/ML/MC/MR/BL/BC/BR]: ***type M***
  6. Specify middle point of text: ***snap to the midpoint of the diagonal line***
  7. Specify height <0.250>: ***1 <enter>***
  8. Specify rotation angle of text <0>: ***0 <enter>***
  9. Enter text: ***type: HHHH <enter>***
 10. Enter text: ***press <enter> to stop***

Also refer to Exercise 8C for "MTP" method.

OTHER JUSTIFICATION OPTIONS:

ALIGN

Aligns the line of text between two points specified.
The height is adjusted automatically.

FIT

Fits the text between two points specified.
The height is specified by you and does not change.

CENTER   HyyHHyyHHyy

This is a tricky one. Center is located at the bottom center of Upper Case letters.

MIDDLE   HHHHHHHHHH   HHyyHHyyHHyy

If only uppercase letters are used: located in the middle, horizontally and vertically.
If both uppercase and lowercase letters are used: located in the middle, horizontally and vertically, of the lowercase letters.

RIGHT   HyyHHyyHHyy

Bottom right of upper case text.

TL, TC, TR   HyyHHyyHHyy

Top left, Top center and Top right of upper and lower case text

ML, MC, MR   HyyHHyyHHyy

Middle left, Middle center and Middle right of upper case text.
(Notice the difference between "Middle" and "MC"

BL, BC, BR   HyyHHyyHHyy

Bottom left, Bottom center and Bottom right of lower case text.

MULTILINE TEXT or MText

MULTILINE TEXT command allows you to easily add a sentence, paragraph or tables. The Mtext editor has most of the text editing features of a word processing program. You can underline, bold, italic, add tabs for indenting, change the font, line spacing, and width of the paragraph.

When using MText you must first define a text boundary box. The text boundary box is defined by entering where you wish to start the text (first corner) and approximately where you want to end the text (opposite corner). It is very similar to drawing a rectangle. The paragraph text is considered one object rather than several individual sentences.

USING MULTILINE TEXT

1. Select the MULTILINE TEXT command using one of the following:

    TYPE = MT
    PULLDOWN = DRAW / TEXT / MULTILINE TEXT
    TOOLBAR = DRAW A

    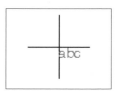

    The command line will lists the current style and text height. The cursor will then appear as crosshairs with the letters "abc" attached. These letters indicate how the text will appear using the current font and text height.

    Mtext Current text style: "STANDARD"   Text height: .250

2. Specify first corner: ***Place the cursor at the upper left corner to start the new text boundary box and press the left mouse button. (P1)***

3. Specify opposite corner or [Height / Justify / Line Spacing / Rotation / Style / Width]: ***Move the cursor to the right and down (P2) and press left mouse button.***

The (2 piece) In-Place Text Editor will appear.

Text Formatting Tool Bar

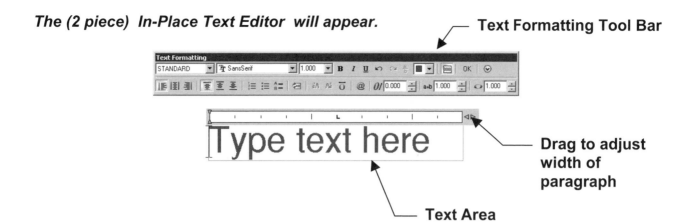

Drag to adjust
width of
paragraph

Text Area

Notice the **In-Place Text Editor** is in 2 pieces: The **Text Formatting toolbar** and the **Text Area**. The toolbar portion can be moved but the text area remains in the location that you designated.

The **Text Formatting toolbar** allows you to select the Text Style, Font, Height etc. You can add features such as bold, italics, underline and color. There is even an UNDO button.

The **Text Area box** allows you to enter the text, add tabs, adjust left hand margins and change the width of the paragraph.

4. After you have entered the text in the Text Area box, do one of the following to add the new text to the drawing and close the Mtext Editor:
   a. Select the OK button
   b. Press Ctrl + <enter>
   c. Left Click anywhere outside the Mtext Editor, but within the drawing area.

HOW TO CHANGE THE "abc", ON THE CROSSHAIRS, TO OTHER LETTERS.

You can personalize the letters that appear attached to the crosshairs using the **MTJIGSTRING** system variable. (10 characters max) The letters will simulate the appearance of the font and height selected but will disappear after you place the lower right corner (P2).

1. Type **MTJIGSTRING** <enter> on the command line.
2. Type the new letters <enter>.

The letters will be saved to the computer, not the drawing. They will appear anytime you use Mtext and will remain until you change them again.

TABS

Setting and removing Tabs is very easy.

The default setting for tabs is 1". (You may set as many tabs as you need.)
Set or change the stop positions at anytime, using one of the following methods.

Method 1.
Place the cursor on the "Ruler" where you want the tab and left click. A little dark "**L**" will appear. The tab is set.
If you would like to remove a tab, just click and drag it off the ruler and it will disappear.

Method 2.
1. Right click on the Ruler.
2. Select "Indents and Tabs" from the short cut menu.

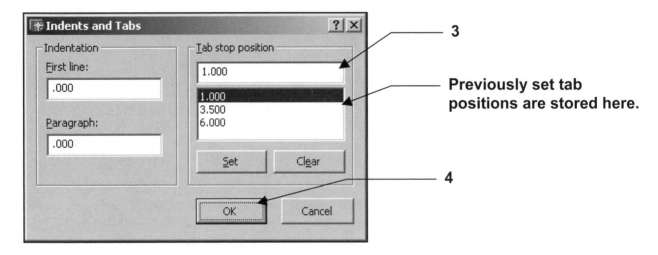

3. Set a tab position by typing the position in the upper box, then select the "Set" button.
4. Select the OK button.

Tab Stop Position
Sets tab positions for the current paragraph or selected paragraphs.
The list below the text box shows the current tab stops.

Set
Copies the value in the Tab Stop Position box to the list below the box.

Clear
Removes the selected tab stop from the list.
Clear a tab position by highlighting the position in the lower box then select the "Clear" button.

INDENTS

Sliders on the ruler show indention relative to the left side of the text boundary box. The top slider indents the first line of the paragraph, and the bottom slider indents the other lines of the paragraph.

You may change their positions at anytime, using one of the following methods.

Method 1.
Place the cursor on the "Slider" and click and drag it to the new location.

Method 2.
1. Right click on the Ruler.
2. Select "Indents and Tabs" from the short cut menu.

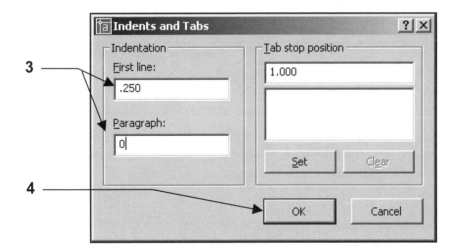

3. Type the indent position in the First line and/or Paragraph box.
4. Select OK button.

First Line:
Sets indentation for the first line of the current paragraph or selected paragraphs.
(The top slider)

Paragraph:
Sets indentation for the current paragraph or selected paragraphs. (The lower slider)

MText – LINE SPACING

The Multiline text command allows you to set the spacing between the bottom of the first line of text to the bottom of the following lines of text. This is accomplished using the **Line spacing** option within the MText command.

You may set the Line spacing to a **Factor** of the "original line spacing", or enter the **Specific** distance desired.

If you choose to use the Factor method, you may use a factor up to 4X the "original line spacing". AutoCAD has established the "original line spacing" as 1.66 of the text height. For example, the line spacing for 1" text is 1.66 from the bottom of the first line to the bottom of the second line.

Line spacing _____
                        (Original line spacing distance is 1.66 X Text ht.)
Line spacing _____

If you choose to enter a specific distance, the original line spacing is ignored.

Set the Line spacing.

1. Select the MText command.

   Command: _mtext Current text style: "STANDARD" Text height: 1.00
2. Specify first corner*: **Place the cursor, and left click, to locate the first corner of the text boundary.**

3. Specify opposite corner or [Height/Justify/Line spacing/Rotation/Style/Width]: **Select the Line spacing option by typing L <enter> or right click and select Line spacing from the short cut menu.**

4. Enter line spacing type [At least/Exactly] <At least>: **Select " Exactly" option by typing E <enter> or right click and select "Exactly" from the short cut menu.** ("At least" is the default option that merely insures that the text does not overlap)

5. Enter line spacing factor or distance <1x>: **Enter a factor or a specific distance. (Note: you must include the "x" when entering the factor number.)**

6. Specify opposite corner or [Height/Justify/Line spacing/Rotation/Style/Width]: **Place the cursor, and click, to locate the opposite corner of the text paragraph.**

   The MText editor will appear.

7. Type the text and select OK.

Edit the Line spacing
To edit the existing line spacing use the Properties Palette.

EDITING TEXT

SINGLE LINE TEXT

Editing **Single Line Text** is somewhat limited compared to Multiline Text. In the example below you will learn how to edit the text within a Single Line Text sentence. (In Lesson 12 you will learn additional options for editing Single Line text by using the Properties command.)

1. Double click on the <u>Single Line text</u> you want to edit. The text will highlight.

2. Make the changes in-place then press <enter><enter>.

MULTILINE TEXT

Multiline Text is as easy to edit as it is to input originally. You may change the style, font, height, color, indent and add text features such as bold, italic and underline.

1. Double click on the Multiline text you want to edit.

The "In-Place Editor" will appear:

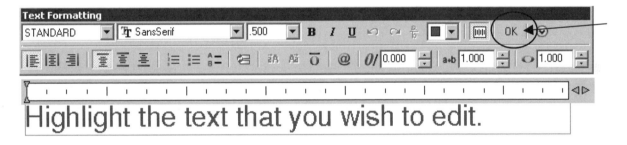

2. Highlight the text, that you want to change, using click and drag.

3. Make the changes then select the **OK** button.

Editing Multiline Text continued….

You may edit many other Multiline Text features by right clicking in the Text Area and selecting an option from the menu shown below.

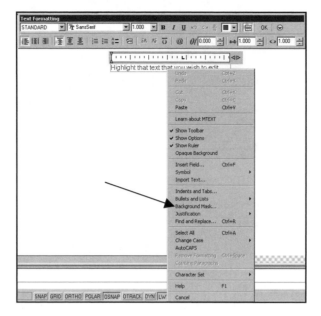

Background Mask

Background mask inserts an opaque background so that objects under the text are covered. (masked) The mask will be a rectangular shape and the size will be controlled by the "Border offset factor". The color can be the same as the drawing background or you may select a different color.

1. After selecting the "Background Mask" option, from the menu shown above, the dialog box will appear.

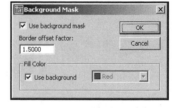

2. Turn this option **ON** by selecting the "Use background Mask" box.

3. Enter a value for the "Border offset factor". The value is a factor of the text height. 1.0 will be exactly the same size as the text. 1.5 (the default) extends the background by 0.5 times the text height. The width will be the same width that you defined for the entire paragraph.

4. In the Fill Color area, select the "Use background" box to make the background the same color as the drawing background. To specify a color, uncheck this box and select a color. (Note: if you use a color you may need to adjust the "draworder" using Tools / Draworder to bring the text to the front.)

SCALING TEXT

The **Scaletext** command allows you to scale <u>Single or Multiline</u> text using **height**, **factor** or **match** a previously drawn text. The text will be scaled proportionately in the X and Y axis. In other words, it gets larger or smaller all over. You must select a "base point" from which the text will enlarge or reduce. The base point is a justification option. If you would like it to scale from the middle, you will select "middle" as the base point. If you would like the text to scale from the top, you will select "TC". Most of the time you will simple use "existing". Existing is the justification used when the text was originally created.

Select the **SCALETEXT** command using one of the following:

> **TYPE = scaletext**
> **PULLDOWN = MODIFY / OBJECT / TEXT / SCALE**
> **TOOLBAR = TEXT**

HEIGHT
1. Select the line of text you want to edit.
2. Select objects: *select more text or <enter> to stop selecting*
3. Enter a base point option for scaling
   [Existing/Left/Center/Middle/Right/TL/TC/TR/ML/MC/MR/BL/BC/BR]
   <Existing>:*select the base point*
4. Specify new height or [Match object/Scale factor] <.250>:*type new height <enter>*

MATCH OBJECT
1. Select the line of text you want to edit.
2. Select objects: *select more text or <enter> to stop selecting*
3. Enter a base point option for scaling
   [Existing/Left/Center/Middle/Right/TL/TC/TR/ML/MC/MR/BL/BC/BR]
   <Existing>:*select the base point*
4. Specify new height or [Match object/Scale factor] <.250>:*select Match object*
5. Select a text object with the desired height: *select the text to match*
   Height = *the new height will be shown here*

SCALE FACTOR
1. Select the line of text you want to edit.
2. Select objects: *select more text or <enter> to stop selecting*
3. Enter a base point option for scaling
   [Existing/Left/Center/Middle/Right/TL/TC/TR/ML/MC/MR/BL/BC/BR]
   <Existing>:*select the base point*
4. Specify new height or [Match object/Scale factor] <.250>:*select Scale factor*
5. Specify scale factor or [Reference] <2.000>: *type the factor*

EXERCISE 8A

INSTRUCTIONS:

1. Start a **New** file and select **1workbook helper.dwt**
2. Duplicate the text shown below using <u>SINGLE LINE</u> text.
3. Use Layer TXT-HVY.
4. Select **DRAW / TEXT / SINGLE LINE Text**
5. Follow the instructions in each block of text. To start the text in the correct location, stated in each example, move your cursor while watching the coordinate display.
6. Save this drawing as EX8A.

DO NOT DUPLICATE NOTES WITH ARROWS

Use justify "Center"
Ht = .50
Center point = 6.25, 8

TEXT EXERCISE

THIS TEXT'S START POINT IS .75, 7.50
AND THE HEIGHT IS .13

THIS TEXT IS JUSTIFIED, RIGHT.
THE ENDPOINT IS 11.50, 7.50.
THE HEIGHT IS .13.
THE TEXT WILL LOOK A LITTLE STRANGE
BECAUSE IT SEEMS TO BE JUSTIFIED LEFT.
BUT WHEN YOU FINISH TYPING AND PRESS
ENTER TWICE, THE TEXT WILL MOVE TO
THE RIGHT JUSTIFIED.

USE JUSTIFY "ALIGN" FOR THIS TEXT. FIRST ENDPT IS .75, 5.5. SECOND ENDPT IS 11.5, 5.5.

USE JUSTIFY "ALIGN" AGAIN. FIRST ENDPT IS .75, 4.75. SECOND ENDPT IS 11.75, 5.25.

USE JUSTIFY "FIT" FOR THIS TEXT.
1ST = 1, 4   SECOND = 5, 4
THE HEIGHT IS .13.
PRESS ENTER AFTER EACH SENTENCE.
THEN THE TEXT WILL JUSTIFY.

22

NO JUSTIFICATION FOR THS TEXT
JUST START TEXT AT 8.5, 1.25
HEIGHT .20 AND ROTATION 45

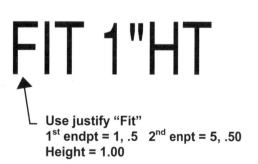
FIT 1"HT

Use justify "Fit"
1st endpt = 1, .5   2nd enpt = 5, .50
Height = 1.00

Draw the Circle first. Center = 7, 3.75
Radius = .75   Text Ht = .50
Use Justify "Middle" to place numbers in the "Center" of the Circle. Numbers will not be in the center until you press enter twice.

EXERCISE 8B

INSTRUCTIONS:
1. Start a **New** file and select **1workbook helper.dwt**
2. Duplicate the text shown below using **MULTILINE** text.
3. Select **DRAW / TEXT / MULTILINE text**
4. Use Text Style: **Standard**
5. Use font: **SansSerif**
6. Text Height: **.250**
7. Follow the directions below.
8. Enter all text shown below.
9. Save as: **EX8B**

The following is a practice exercise for tabs, indent and the features bold and underline.
    1. This sentence should be indented 1 inch.
        a. This sentence should be indented 1/2 inch more.
And now back to the left margin.

Isn't this fun!

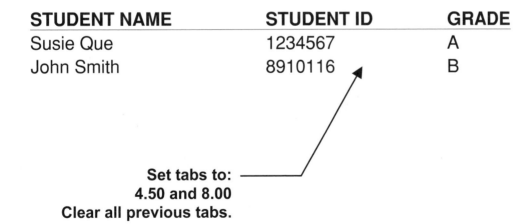

STUDENT NAME	STUDENT ID	GRADE
Susie Que	1234567	A
John Smith	8910116	B

Set tabs to:
4.50 and 8.00
Clear all previous tabs.

EXERCISE 8C

INSTRUCTIONS:

1. Start a **New** file and select **1workbook helper.dwt**

The following exercise is designed to teach you how to insert text into the exact middle of a rectangular area. You will use both <u>Single Line Text</u> and <u>Multiline Text</u>.

2. Follow the instructions for each text box below.
3. Save as: **EX8C**

SINGLE LINE TEXT

1. Draw a 6" wide by 3" high rectangle.
2. Select "**SINGLELINE TEXT**"
3. Use Justify: Middle
4. Use "**MTP**" object snap to locate the middle of the rectangle.
    a. Type **MTP** <enter> on the command line.
    b. Using object snap "Endpoint" snap to **(P1)** corner and then the diagonal corner **(P2)**
5. Use Text Ht: 1"
6. Rotation "0"
7. Type the word "Middle" and <enter><enter>.

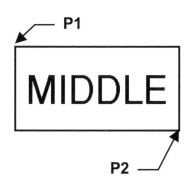

MULTILINE TEXT

1. Draw a 13" wide by 2" high rectangle.
2. Select "**MULTILINE TEXT**"
3. Start the Text boundary box at the upper left corner **(P1)** of the rectangle. Place the opposite corner at the lower right corner **(P2)** of the rectangle. (Use "Endpoint" object snap to be accurate.)
4. Select **Center, Middle, Standard, Sans Serif, 1.00**

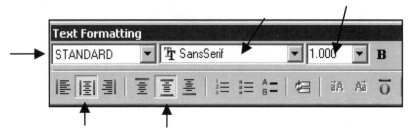

5. Type "**MIDDLE CENTER**" (<u>don't press</u> <enter>) select **OK.**

EXERCISE 8D

INSTRUCTIONS:

1. Start a **New** file and select **1workbook helper.dwt**
2. Draw **two** 6" long lines as shown.
3. Select Draw / Text / Single Line Text
   a. Select **Justify - Center**.
   b. Use Midpoint snap to place the justification point at the midpoint of the line.
   c. Use text height 1" and rotation angle 0.
   d. Type the word "Happy" <enter> <enter> (Use upper and lower case)

4. Select Draw / Text / Single Line Text again.
   a. This time select **Justify - BC**. (bottom center)
   b. Use Midpoint snap to place the justification point at the midpoint of the line.
   c. Use text height 1" and rotation angle 0.
   d. Type the word "Happy" <enter> <enter> (Use upper and lower case)

Notice the difference between <u>Center</u> and <u>Bottom Center</u>?

"Center" only considers the Upper Case letters when justifying.
"Bottom Center" is concerned about those Lower Case letters.
Can you see how you could accidentally place your text too high or too low? Think about the difference between Center and Bottom Center.

5. Save as: **EX8D**

EXERCISE 8E

INSTRUCTIONS:

1. Start a **New** file and select **1workbook helper.dwt**
2. Select Draw / Text / Multiline Text
3. Specify the size of the multiline paragraph by entering the coordinates for the first corner and then the coordinates for the opposite corner exactly as follows:
    First corner Type: 7,7<enter>   Opposite Corner Type: 14, 10 <enter>
3. Use the following:
    Text style: Standard   Font: SansSerif   Height: .250
4. Type the following as **one continuous line**. ***Do not press <enter>.***
5. When you are finished typing, select OK.

Today is the day that I will learn how to use
AutoCAD's text editing features. I will be a
good student and practice **all** the features
because I **really** want to learn!

Hopefully your text appears the same as the text shown above.

Now you are going to change the width of this paragraph.

5. Select the in-place editor by double clicking on the paragraph.
6. Reduce the "ruler" to 4 inches as follows.
    a. Move the cursor to the ruler and press the right mouse button.
    b. Select "Set Mtext width"
    c. Enter **4** and **OK**
7. Select OK.

Today is the day that I will
learn how to use
AutoCAD's text editing
features. I will be a good
student and practice **all**
the features because I
really want to learn!

Does your text look like the example above?

8. Save as: **EX8E**

EXERCISE 8F

INSTRUCTIONS:

1. Start a **New** file and select **1workbook helper.dwt**.
2. Draw the Text on the left using "Single Line" text option for the first sentence and "Multiline" text option for the second sentence. (Layer = Txt-Hvy)
3. Type **Mirrtext** at the command line.
4. Select setting **1** (this is the default setting)
5. Using the **Mirror** command, mirror both sentences using a vertical mirror line, **P1** to **P2**, approximately as shown.

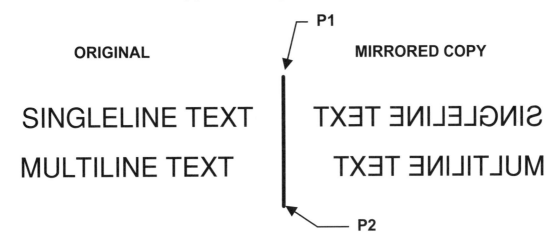

6. Now change the **Mirrtext** setting to **0**
7. Try the Mirror command again.

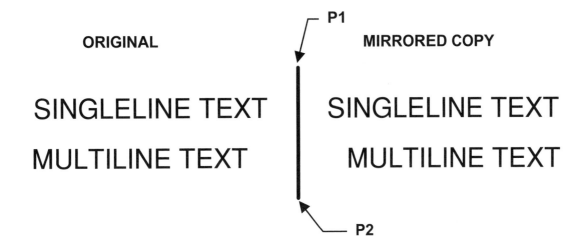

8. Save this drawing as **EX8F**.

NOTES:

LEARNING OBJECTIVES

After completing this lesson, you will be able to:

1. Understand the ORIGIN.
2. Draw Objects using Coordinate Input.
3. Input Absolute and Relative coordinates.
4. Use Direct Distance Entry.
5. LIST information about objects.
6. Determine the distance between two points.
7. Identify a location within the drawing.
8. Create your own 11 x 17 Master Border.
9. Print 11 x 17 drawings in Model Space.

LESSON 9

COORDINATE INPUT

In the previous lessons you have been using the cursor to place objects.
In this lesson you will learn how to place objects in <u>specific locations</u> by entering coordinates. This process is called **Coordinate Input**.

This is not difficult, so do not start to worry.

Autocad uses the ***Cartesian Coordinate System.***

The Cartesian Coordinate System has 3 axes, X, Y and Z.

The **X** is the Horizontal axis. *(Right and Left)*
The **Y** is the Vertical axis. *(Up and Down)*
The **Z** is Perpendicular to the X and Y plane.
*(The **Z** axis, <u>which is not discussed in this workbook</u>, is used for 3D.)*

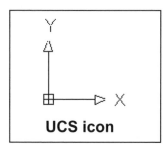

UCS icon

Look at the User Coordinate System (UCS) icon in the lower left corner of your screen.
The arrows are pointing in the positive direction.

The location where the X , Y and Z axes intersect is called the **ORIGIN**. (0,0,0)
Currently the Origin is located in the lower left corner of the screen.
When you move the cursor away from the Origin, in the direction of the arrows, the X and Y coordinates are positive.
When you move the cursor in the opposite direction, the X and Y coordinates are negative.

Using this system, every point on the
screen can be specified using positive
and negative X and Y coordinates.

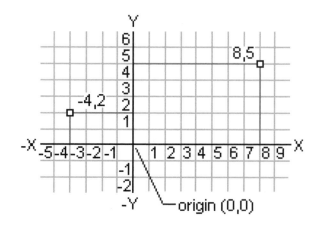

**Note: Please confirm that
Dynamic Input is "Off" by releasing
the "DYN" button in the status bar
or press F12.
Dynamic Input will be discussed in
Lesson 11.**

There are 3 types of Coordinate input, **Absolute**, **Relative** and **Polar**.
(Polar will be discussed in Lesson 11)

ABSOLUTE COORDINATES   (The input format is: **X, Y**)

Absolute coordinates come *from the ORIGIN* and are typed as follows: **8, 5 .**
The first number (8) represents the **X-axis** (horizontal) distance <u>from the Origin</u> and the
second number (5) represents the **Y-axis** (vertical) distance from the Origin.
The two numbers must be separated by a **comma**.

An absolute coordinate of **4, 2** will be **4** units to the right (horizontal) and **2** units up
(vertical) <u>from the current location of the Origin.</u>

An absolute coordinate of **-4, -2** will be **4** units to the left (horizontal) and **2** units down
(vertical) <u>from the current location of the Origin.</u>

The following are examples of Absolute Coordinate input.
<u>Notice where the Origin is located in each example.</u>

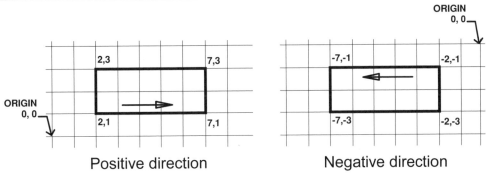

Positive direction   Negative direction

RELATIVE COORDINATES (The input format is: **@X, Y**)

Relative coordinates come *from the last point entered*. The first number represents
the **X-axis** (horizontal) and the second number represents the **Y-axis** (vertical).
The two numbers must be separated by a **comma**. To distinguish between Absolute
and Relative, use the **@** symbol and then the X and Y coordinates.

A Relative coordinate of **@5, 2** will go to the **right** 5 units and **up** 2 units
<u>from the last point entered.</u>

A Relative coordinate of **@-5, -2** will go to the **left** 5 units and **down** 2 units
<u>from the last point entered.</u>

The following is an example of Relative Coordinate input.

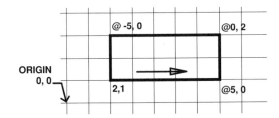

EXAMPLES OF COORDINATE INPUT

Scenario 1.
You want to draw a line with the first endpoint "at the Origin" and the second endpoint 3 units in the positive X direction.

1. Select the Line command.
2. You are prompted for the first endpoint: **Type 0, 0 <enter>**
3. You are then prompted for the second endpoint: **Type 3, 0 <enter>**

What did we do?
The first endpoint coordinate input, 0,0 means that you do not want to move away from the Origin. You want to start "on" the Origin.

The second endpoint coordinate input, 3, 0 means that you want to move 3 units in the positive X axis. The "0" means you do not want to move in the Y axis. So the line will be exactly horizontal.

Scenario 2.
You want to start a line 8 units directly above the origin and it will be 4 units in length, perfectly vertical.

1. Select the Line command.
2. You are prompted for the first endpoint: **Type 0, 8 <enter>**
3. You are prompted for the second endpoint: **Type @0, 4 <enter>**

What did we do?
The first endpoint coordinate input, 0, 8 means you do not want to move in the X axis direction but you do want to move in the Y axis direction.

The second endpoint coordinate input @0, 4 means you do not want to move in the X axis "from the last point entered" but you do want to move in the Y axis "from the last point entered. (Remember the @ symbol tells AutoCAD that the coordinates typed are relative to the "last point entered" not the Origin.

Scenario 3.
Now try drawing 5 connecting lines sements.

1. Select the Line command.
2. First endpoint: 2, 4 <enter>
3. Second endpoint: @ 2, -3 <enter>
4. Second endpoint: @ 0, -1 <enter>
5. Second endpoint: @ -1, 0 <enter>
6. Second endpoint: @ -2, 2 <enter>
7. Second endpoint: @ 0, 2 <enter> <enter>

Notice the "@" symbol for relative coordinates.

Note: If you enter an incorrect coordinate, just type "U" <enter> and the last segment will disappear and you will have another chance at entering the correct coordinate.

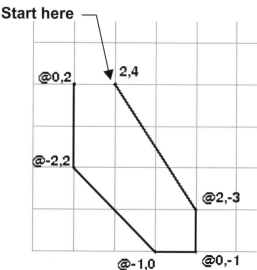

Start here

@0,2   2,4

@-2,2

@2,-3

@-1,0   @0,-1

DIRECT DISTANCE ENTRY (DDE)

F8 - Ortho

DIRECT **DISTANCE ENTRY** is a combination of keyboard entry and cursor movement. **DDE** is used to specify distances in the horizontal or vertical axes <u>from the last point entered</u>. **DDE** is a ***Relative Input.*** Since it is used for Horizontal and Vertical movements, **Ortho** must be **ON**.

(Note: to specify distances on an angle, refer to Lesson 11)

**Using DDE is simple. Just move the cursor and type the distance.
Negative and positive is understood automatically by moving the cursor up (positive), down (negative), right (positive) or left (negative) from the last point entered. No minus sign necessary.**

Moving the cursor to the right and typing 5 and <enter> tells AutoCAD that the 5 is positive and Horizontal.
Moving the cursor to the left and typing 5 and <enter> tells AutoCAD that the 5 is negative and Horizontal.
Moving the cursor up and typing 5 and <enter> tells AutoCAD that the 5 is positive and Vertical.
Moving the cursor down and typing 5 and <enter> tells AutoCAD that the 5 is negative and Vertical.

EXAMPLE:

1. <u>Ortho must be ON</u>.
2. Select the Line command.
3. Type: 1, 2 <enter> to enter the first endpoint using Absolute coordinates.
4. Now move your cursor to the right and type: 5 <enter>
5. Now move your cursor up and type: 4 <enter>
6. Now move your cursor to the left and type: 5 <enter> <enter> to stop

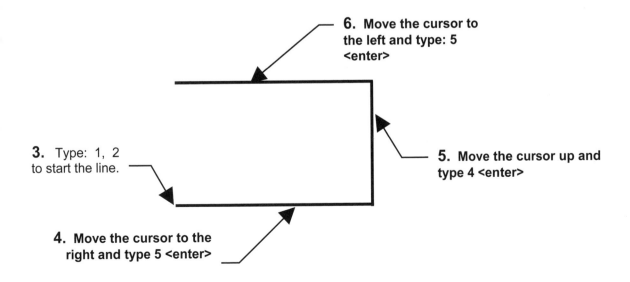

6. Move the cursor to the left and type: 5 <enter>

3. Type: 1, 2 to start the line.

5. Move the cursor up and type 4 <enter>

4. Move the cursor to the right and type 5 <enter>

INQUIRY — Tools/Inquiry

The INQUIRY command allows you to Inquire about objects on the screen. There are 8 commands in the inquiry menu but we will only discuss 3 of those commands at this time. The remaining 5 commands will be discussed in the "Advanced" workbook.

LIST

The LIST command will list the type of object you have selected, coordinate location and properties that apply to the object.

1. Select the LIST command using one of the following:

TYPE = LIST
PULLDOWN = TOOLS / INQUIRY / List
TOOLBAR = STANDARD

2. Select the object: *select the object*
3. Select the object: *press <enter> to stop*
(The information will be listed in the "Text Screen". Press F2 to close the Text Screen.)

DISTANCE

The DISTANCE command will list the distance between two points that you select.

1. Select the DISTANCE command using one of the following:

TYPE = DI
PULLDOWN = TOOLS / INQUIRY / Distance
TOOLBAR = STANDARD

2. First point: *select the first point*
3. Second point: *select the second point*
   Distance = *distance between the two points will be listed here*

ID POINT

The ID POINT or LOCATE POINT command will list the X and Y coordinates of the point that you selected. The <u>coordinates will be from the ORIGIN</u>.

1. Select the ID POINT command

TYPE = ID
PULLDOWN = TOOLS / INQUIRY / ID Point
TOOLBAR = TOOLS

2. Point: *select a point anywhere on the drawing*
   X = *coordinate listed here* Y = *coordinate listed here* Z = *coordinate listed here*

Note: the **ID POINT** can also be used to create a **LAST POINT**. This enables you to use Relative coordinates (the @ symbol) for the location of the next object.

LINEWEIGHTS

You learned in Lesson 3 (page 3-8) that it is "good drawing management" to draw related objects on the same layer. It is also "good drawing management" to establish a contrast in line weights between layers. For example, objects such as a house or a paper clip should be drawn with the "Object" layer and should have a greater line weight than the dimension layer or text layer.

The following are instructions for assigning Lineweights to Layers.

FIRST YOU NEED TO CHANGE THE LINEWEIGHT SETTINGS BOX.
1. Select **Format / Lineweight.**

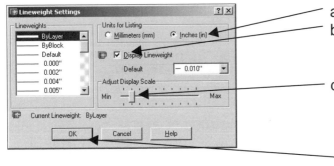

a. Select Inches or millimeters.

b. Select "Display Lineweight" box (you may use status line button LWT)

c. Slide "Adjust Display Scale" to the left as shown. (Controls Lineweight appearance on the screen only)

d. Select OK

(These settings will be saved to the computer not the drawing and will remain until you change them.)

ASSIGNING LINEWEIGHTS TO LAYERS
2. Select **Format / Layer**

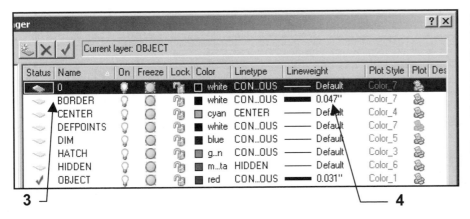

3. ⌐

4.

Notice the 0.047" Lineweight setting for the "Border" layer and the 0.031" Lineweight setting for the "Object" layer.

Default = 0.010

3. Select the "Border" layer. (Click on the name "Border")

4. Click on the Lineweight.

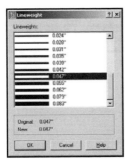

5. Scroll and select from the list, then OK.

*(Lineweight changes will be saved **only to the current drawing**, if you save it, and will not affect any other drawing)*

EXERCISE 9A

INSTRUCTIONS:

explode rectangle

1. Open **9A Helper.dwg**
(If you do not have this drawing
refer to page 2-2)

offset

2. Set **UNITS** using:
    Format / Units
    Units = Decimal
    Precision = 0.000

*tr = trim erase
lines that you
don't want*

3. Set **DRAWING LIMITS** using:
    Format / Drawing Limits
    Lower Left Corner = 0,0
    Upper Rt. Corner = 17, 11

4. Show the new limits using:
    View / Zoom / All

5. Change Lineweight settings to <u>Inches</u>
    and adjust <u>Display scale</u>. (See pg. 9-7)

6. Set **GRIDS** and **SNAP** using:
    Tools / Drafting Settings
        Snap = ON = .250
        Grids = ON = .500

7. **Draw the border below** to scale using
    the dimensions shown.
    Use layer **BORDER**.

8. Place the **TEXT** as shown.
    Use "**Single Line Text**"
    Use Layer = **Text-Hvy**
    Text Ht. = .25

9. **Save** this border as **BSIZE**.

10. **PLOT** using:
    File / Plot
    *Refer to "Basic Plotting from
    Model Space" on the next page.*

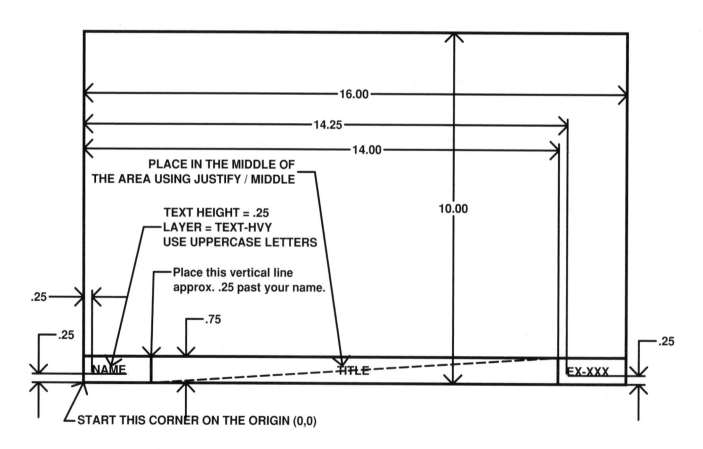

BASIC PLOTTING FROM MODEL SPACE

Note: More Advanced plotting methods will be explained in Lessons 26 and 27.

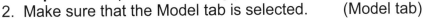

1. **Important:** Open the drawing you want to plot.
2. Make sure that the Model tab is selected.   (Model tab)
3. Select: View / Zoom / All
4. Select the **Plot** command by "right clicking" on the "Model" tab or using one of the following methods listed below:

> **Type = Print or Plot**
> **Pulldown = File / Plot**
> **Tool bar = Standard**

The Plot dialog box below should appear.

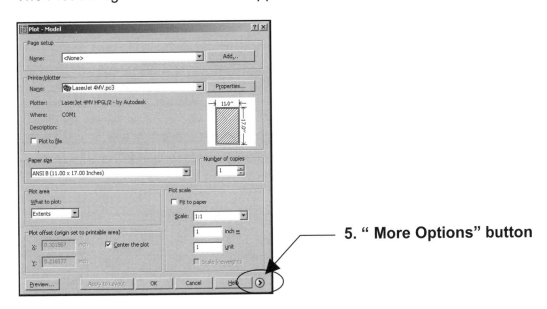

5. " More Options" button

5. Select the "**More Options**" button to expand the dialog box.

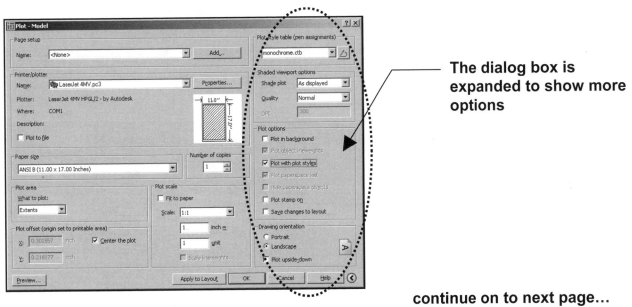

The dialog box is expanded to show more options

continue on to next page…

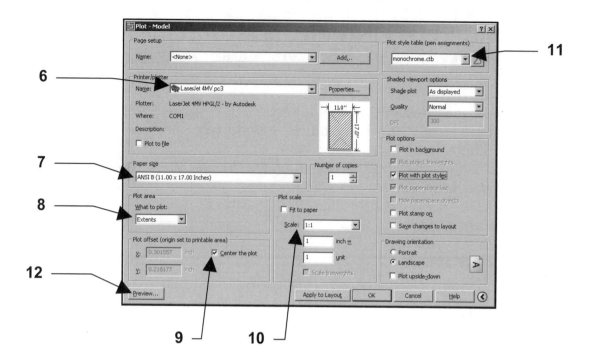

6. Select the "Printer / Plotter" **LaserJet 4MV** from the drop down list.
   Note: This Printer represents a size 17 X 11 for the exercises in this workbook. If it is not in the drop down list, you need to configure it on your system, even though you do not have this printer. Don't worry, this will not harm your computer. Refer to Appendix A for step by step instructions.
   (If you would like to use your own 8-1/2 X 11 printer, you may select it but refer to the note about scale selection in #10 below)

7. Select the Paper Size **ANSI B (11 X 17 inches) (See #10 below)**

8. Select the Plot Area **EXTENTS**

9. Select the Plot Offset **Center the plot**

10. Select Plot Scale **1 : 1**
    (Note: If you would like to print your drawing on a 8-1/2 X 11 printer, select the printer and change the scale to "Fit to Paper" and change paper size. But first configure your drawing for the LaserJet 4MV to match the exercises in the workbook)

11. Select the Plot Style table **Monochrome.ctb**
    The following box will appear. Select **Yes**

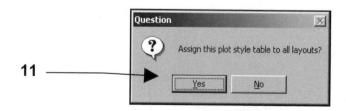

Continue on to next page...

12. Select **Preview** button.

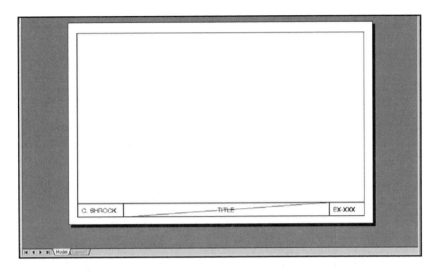

Does your display appear as shown above?
If yes, press <enter> and proceed to 13.
If no, recheck 1 through 11.

You have just created a **Page Setup**. All of the settings you have selected can now be saved. You will be able to recall these settings for future plots using this page setup.
To save the **Page Setup** you need to **ADD** it to the model tab within this drawing.

13. Select the ADD button.

13

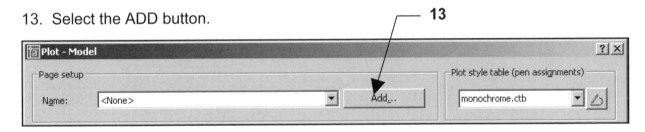

14. Type the new Page Setup name **Model-mono**
(This name identifies that you will use it when plotting the Model tab in monochorme.)

15. Select **OK** button

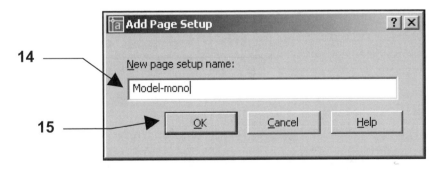

continue on to next page....

16. Select **Apply to Layout** button.

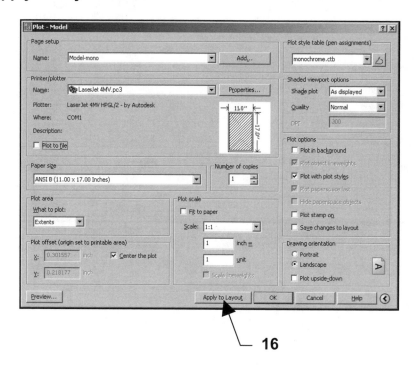

16

17. Select the **OK** button to send the drawing to the printer or select **Cancel** if you do not want to print the drawing at this time. The Page Setup will still be saved.

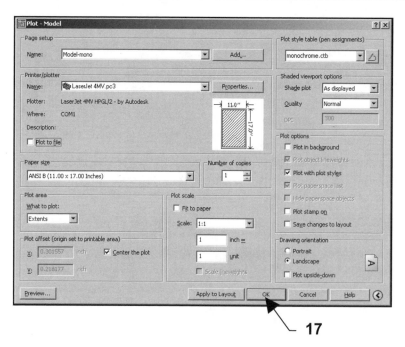

17

18. Save the entire drawing again. The Page Setup will be saved to the **Model tab** within the drawing and available to select in the future. You will not have to select all the individual settings.

Note: The instructions above are to be used when plotting from the "Model" tab. In lessons 26 and 27 you will learn how to plot in a "Layout" or what is commonly referred to as Paperspace. But you have much more to learn first. Let's just take it step by step and not get confused. You will learn it all by the end of this workbook.

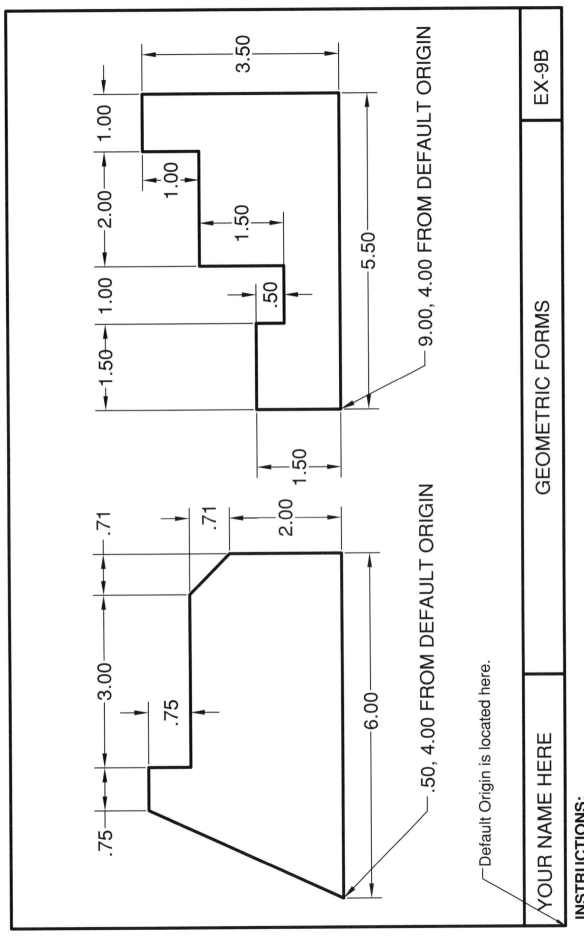

GEOMETRIC FORMS

EX-9B

YOUR NAME HERE

EXERCISE 9B

3.50

1.00

1.00

1.50

.50

2.00

1.00

1.50

5.50

9.00, 4.00 FROM DEFAULT ORIGIN

.71

.71

2.00

3.00

.75

.75

6.00

.50, 4.00 FROM DEFAULT ORIGIN

Default Origin is located here.

INSTRUCTIONS:
1. Open border **BSIZE.**
2. Draw the objects above using Absolute and Relative coordinates.
3. Use Layer = **Object.**
4. Edit the Title Block text by double clicking on the text. <u>Do not erase and replace!</u>
5. Do not dimension.
6. Save as **EX9B** and **Plot using Page Setup "Model-Mono" (Pg. 9-11).**

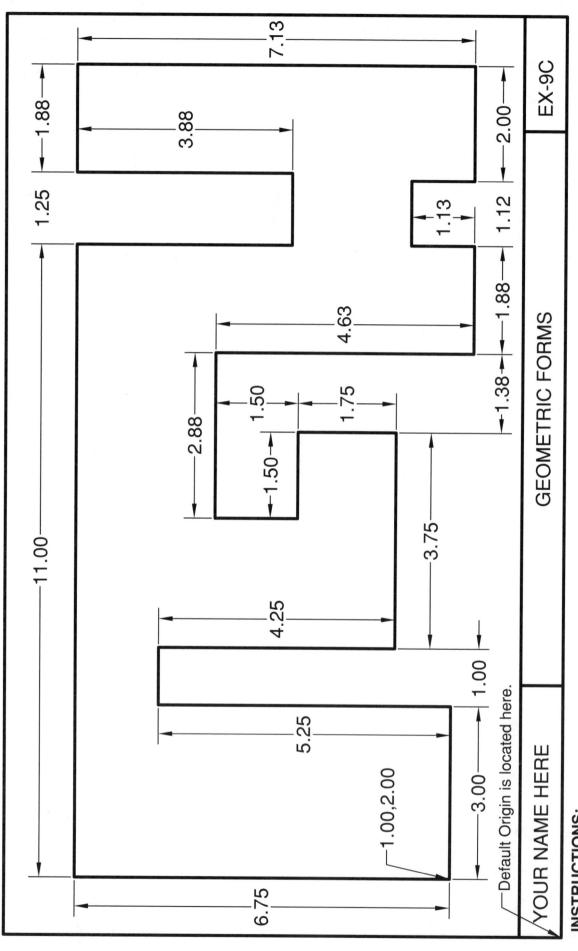

EXERCISE 9C

GEOMETRIC FORMS

EX-9C

YOUR NAME HERE

INSTRUCTIONS:

1. Open border **BSIZE.**
2. Draw the objects above using Absolute and Relative coordinates.
3. Use Layer = **Object.**
4. Edit the Title Block text by double clicking on the text. Do not erase and replace!
5. Do not dimension.
6. Save as **EX9C** and **Plot** using **Page Setup "Model-Mono" (Pg. 9-11).**

9-14

LEARNING OBJECTIVES

After completing this lesson, you will be able to:

1. Move the Origin.
2. Turn the UCS Icon On and Off.
3. Command the UCS Icon to move with the Origin.

LESSON 10

MOVING THE ORIGIN

As previously stated in Lesson 9, the **ORIGIN** is where the X, Y, and Z axes intersect. The Origin's (0,0,0) default location is in the lower left-hand corner of the drawing. But you can move the Origin anywhere on the screen using the UCS command.
(The default location is designated as the "**World**" option or WCS. When it is moved it is UCS, User Coordinate System.)

You may move the Origin many times while creating a drawing. This is not difficult and will make it much easier to draw objects in specific locations. You will understand this better after completing 10A, 10B and 10C.

To MOVE the Origin:

1. Select one of following:

> **TYPING = none**
> **PULLDOWNS = TOOLS / New UCS / Origin**
> **UCS II TOOLBAR =**

Command: _ucs
Current ucs name: *World*
Specify origin of UCS or [Face/NAmed/OBject/Previous/View/World/X/Y/Z/ZAxis] <World>: _o

2. Specify new origin point <0,0,0>: *type coordinate or use the cursor to place.*

To RETURN the Origin to the default "World" location (the lower left corner):

1. Select one of the following:

> **TYPING =   UCS <enter> W <enter>**
> **PULLDOWNS = TOOLS / New UCS / World**
> **UCS TOOLBAR =**

DISPLAYING THE UCS ICON

The UCS icon is merely a drawing aid. You control how it is displayed.
It can be visible (on) or invisible (off). It can move with the Origin or stay in the default location. You can even change it's appearance.

Select the following pull-down menu:

View / Display / UCS Icon

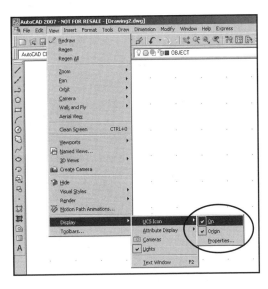

ON: A check mark beside the word ON means the UCS icon will be visible.
Remove the check mark and the UCS icon will disappear.
It will be very helpful to have the UCS Icon "On" most of the time. It displays where the Origin is located. (Set to ON for the workbook exercises)

ORIGIN: A check mark beside the word Origin will move the UCS Icon with the Origin each time you move the Origin. Remove the check mark and the icon will not move. *I find it very helpful to know where the Origin is at all times by merely looking for the UCS icon.* (Set to ON for the workbook exercises)

PROPERTIES: This setting allows you to change the appearance of the UCS icon.
When you select this option the dialog box shown below will appear.
You may change the Style, Size and Color at any time. Changing the appearance is personal preference and will not affect the drawing or commands.

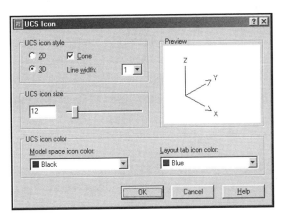

EXERCISE 10A

NOTE: USE LINES NOT POLYGON

PLACE AT CENTER OF DRAWING AREA USING MIDPOINT SNAP.

4.340

1.640

Ø1.000

2.330

2.500

5.000

Move the Origin here
before drawing the Circle.

1.000

2. Move the Origin here
before drawing the
triangular shape above.

RECTANGLE AND LINES

EX-10A

3.000

1.250

4.000

1.750

Ø1.250

3.000

Move the Origin
here before
drawing the Circle.

1.875

2. Move Origin here
before drawing the
rectangle above.

NAME

INSTRUCTIONS:

1. Open **BSIZE**.
2. First move the Origin to the location noted, then input coordinates for drawing.
3. Use Layer: **OBJECT**
4. Do not dimension.
5. Edit the Title Block text.
6. Save as **EX10A** and **Plot**.

10-4

Notice, by moving the Origin, you do not have to add or subtract dimensions. You merely type the X and Y coordinates.

4X Ø.750

Ø2.000

2.500

.750

4.000

1.000

6.000

3.000

3. Move the Origin here before you draw the 3 remaining small circles.

2. Move the Origin here before drawing the large Circle and the small lower left circle.

2.000

4.000

2.500

4.750

1. Move the Origin here before drawing the rectangle above

WINDOW HINGE PLATE

NAME

EX-10B

EXERCISE 10B

INSTRUCTIONS:

1. Open **BSIZE**.
2. First move the Origin to the location noted, then input coordinates for drawing.
3. Use Layer: **OBJECT**
4. Do not dimension.
5. Edit the Title Block text.
6. Save as **EX10B** and **Plot**.

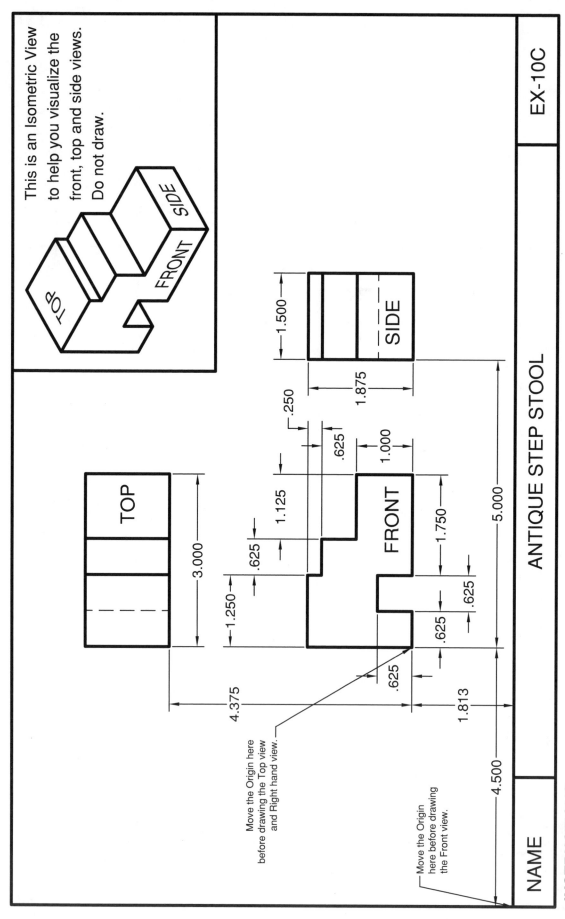

This is an Isometric View to help you visualize the front, top and side views. Do not draw.

TOP

FRONT

SIDE

TOP

3.000

SIDE

1.500

1.875

.250

.625

1.000

FRONT

1.125

.625

1.250

1.750

.625

.625

5.000

.625

1.813

4.375

4.500

Move the Origin here before drawing the Top view and Right hand view.

Move the Origin here before drawing the Front view.

EX-10C

ANTIQUE STEP STOOL

EXERCISE 10C

NAME

INSTRUCTIONS:
1. Open **BSIZE**.
2. First move the origin, then input coordinates. Try using DDE (ref. 9-5).
3. Do not dimension.
4. Use Layer: **OBJECT** and **HIDDEN** (for hidden lines).
5. Edit the Title Block text.
6. Save as **EX10C** and **Plot**.

LEARNING OBJECTIVES

After completing this lesson, you will be able to:

1. Understand the Polar Degree Clock.
2. Draw Lines to a specific length and angle
3. Draw Objects using Polar Coordinate Input.
4. Use Dynamic Input.
5. Construct an Isometric view.

LESSON 11

POLAR COORDINATE INPUT

In Lesson 9 you learned to control the length and direction of horizontal and vertical lines using Relative Input and Direct Distance Entry. Now you will learn how to control the length and **ANGLE** of a line using **POLAR COORDINATE INPUT**.

UNDERSTANDING THE *"POLAR DEGREE CLOCK"*

Previously when drawing Horizontal and Vertical lines you controlled the direction using a <u>Positive</u> or <u>Negative</u> input. ***Polar Input is different***. The Angle of the line will determine the direction. For example: If I want to draw a line at a 45 degree angle towards the upper right corner, I would use the angle 45. But if I want to draw a line at a 45 degree angle towards the lower left corner, I would use the angle 225. You may also use Polar Input for Horizontal and Vertical lines using the angles 0, 90, 180 and 270. No negative input is required.

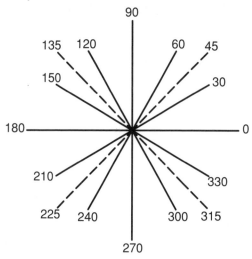

DRAWING WITH *POLAR COORDINATE INPUT*

A Polar coordinate may come ***from the last point entered*** or ***from the Origin,*** depending upon whether you use the @ symbol or not. The first number represents the **Distance** and the second number represents the **Angle**. The two numbers are separated by the **less than (<) symbol**.
The input format is: **distance < angle**

A Polar coordinate of **@6<45** will be 6 units long and at an angle of 45 degrees ***from the last point entered.***

A Polar coordinate of **6<45** will be 6 units and 45 degrees from the ***Origin***.

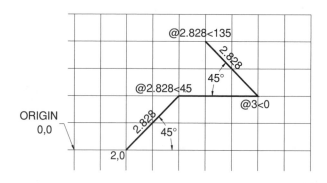

USING POLAR COORDINATES and DDE

The following is a simple drawing to practice typing Polar coordinates and Direct Distance Entry. After you have completed this lesson you may decide which method of input you prefer.

       1 Set the Status Bar as follows:
         Ortho = ON   All others = OFF

       2. Select the **Line** command:

       3. Start the Line anywhere in the drawing area.

Line A   4. Move the cursor to the right.
           5. Type 2 <enter>

Line B   6. Type @3 < 45 <enter>

Line C   7. Move the cursor up.
           8. Type 2 <enter>

Line D   9. Type @4 < 225 <enter> (Refer to the degree clock on page 11-2)

Line E   10. Move the cursor to the left.
           11. Type 1.293 <enter>

Line F   12. Move the cursor down.
           13. Type 1.293 <enter>

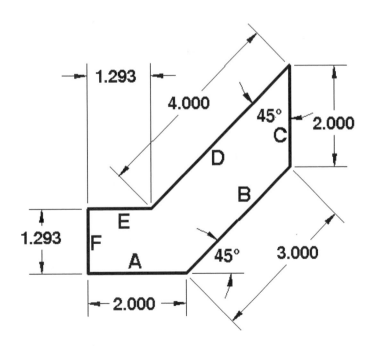

DYNAMIC INPUT

To help you keep your focus in the "drawing area", AutoCAD has provided a command interface called **Dynamic Input**.

When Dynamic Input is ON abbreviated prompts that you normally see on the command line are displayed near the cursor. You may also input information within the Dynamic Input instead of on the command line.

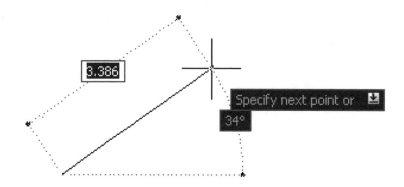

Note: *Some users find Dynamic Input useful some find it distracting. After completing this lesson you decide if you want to use it. __It is your choice__.*

How to turn Dynamic Input ON or OFF

Select the **DYN** button on the status bar or use the F12 key.

DYNAMIC INPUT has 3 components

1. **Pointer Input**
2. **Dimensional Input**
3. **Dynamic Prompts**

You may control what is displayed by each component and turn each ON or OFF.

Refer to the following pages for more detailed descriptions.

POINTER INPUT

Pointer Input is only displayed for the **first** point.
When Pointer Input is enabled (ON) and a command has been selected, the location of the crosshairs is displayed as coordinates in a tooltip near the cursor.

Example:
1. Select the **LINE** command.

2. Move the cursor.

The 2 boxes display the cursor location as X and Y coordinates **from the Origin**. (Absolute coordinates)

3. You may move the cursor until these coordinates display the desired location and press the left mouse button or enter the desired coordinate values in the tooltip boxes instead of on the command line.

> **Note:**
> Refer to "**How to enter coordinate values in Dynamic Input tooltips**" page 11-8.

How to change POINTER INPUT settings

1. Right-click the **Dyn** button on the status bar and select **Settings**.

2. In the Drafting Settings dialog box select the **Dynamic input** tab.

3. Under Pointer Input select the **Settings** button.

4. In the Pointer Input Settings dialog box select:

   Format: (select one as the display format)

   > **Polar** or **Cartesian** format

   > **Relative** or **Absolute** coordinate

   Visibility: (select one for tooltip display)

   - **As Soon As I Type Coordinate Data**. When pointer input is turned on, displays tooltips only when you start to enter coordinate data.

   - **When a Command Asks for a Point**. When pointer input is turned on, displays tooltips whenever a command prompts for a point.

   - **Always—Even When Not in a Command**. Always displays tooltips when pointer input is turned on.

5. Select **OK** to close each dialog box.

DIMENSIONAL INPUT

When Dimensional Input is enabled (ON) the tooltips display the distance and angle values for the **second** and **subsequent points**.

The default display is Relative Polar coordinates.

Dimensional Input is available for: **Arc, Circle, Ellipse, Line** and **Pline**.

Example:
1. Select the Line Command.

2. Enter the 1st point.

3. You may enter the **distance** and then the **angle** in the tooltip boxes instead of on the command line.

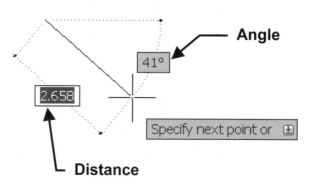

Note:
Refer to "How to enter coordinate values in Dynamic Input tooltips" page 11-8.

How to change DIMENSIONAL INPUT settings

1. Right-click the **Dyn** button on the status bar and select **Settings**.

2. In the Drafting Settings dialog box select the **Dynamic input** tab.

3. Under Dimension Input, select the **Settings** button.

4. In the Dimension Input Settings dialog box select:

    Visibility: (select one of the following options)

    o **Show Only 1 Dimension Input Field at a Time**. Displays only the distance dimensional input tooltip when you are using grip editing to stretch an object.

    o **Show 2 Dimension Input Fields at a Time**. Displays the distance and angle dimensional input tooltips when you are using grip editing to stretch an object.

    o **Show the Following Dimension Input Fields Simultaneously**. Displays the selected dimensional input tooltips when you are using grip editing to stretch an object. Select one or more of the check boxes.

5. Select **OK** to close each dialog box.

DYNAMIC PROMPTS

When Dynamic Prompts are enabled (ON) prompts are displayed.

You may enter a response in the tooltip box instead of on the command line.

Example:

1. Select the <u>Circle</u> command.

2. Place the <u>center point</u> for the circle by entering the X and Y coordinate values in the tooltip boxes or press the left mouse button.

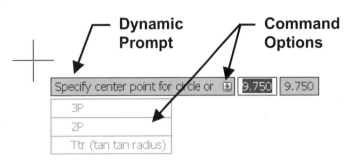

3. Press the right mouse button for command options or you may also press the down arrow to view command options.
   Select an option by clicking on it or press the up arrow for option menu to disappear.

4. Enter Radius or select Diameter using one of the methods in 3 above.

How to change the COLOR, SIZE, or TRANSPARENCY of tooltips

1. Right-click the **Dyn** button on the status bar then select **Settings**.

2. In the Drafting Settings dialog box select the **Dynamic input** tab.

3. At the bottom of the dialog box select **Drafting Tooltip Appearance** button.

4. Under Color, select the Model Color or Layout Color button to display the **Select Color** dialog box. You may specify a color for tooltips in the space you selected.

   (I like color 9, a light gray.)

5. Under **Size**, move the slider to the right to make tooltips larger or to the left to make them smaller. The default value, 0, is in the middle.

   (I like 2, a little larger than the default)

6. Under **Transparency**, move the slider. The higher the setting, the more transparent the tooltip.

   (I like 25, a little more transparent than the default)

7. Under **Apply To**, choose an option:

   o **Override OS Settings for All Drafting Tooltips.** Applies the settings to all tooltips, overriding the settings in the operating system.

   o **Use Settings Only for Dynamic Input Tooltips.** Applies the settings only to the drafting tooltips used in Dynamic Input.

8. Select **OK** to close each dialog box.

HOW TO ENTER COORDINATE VALUES IN DYNAMIC INPUT TOOLTIP BOXES

Use one of the following methods.

To enter Cartesian coordinates (X and Y)

1. Enter an "**X**" coordinate value <u>and a **comma**</u>.

2. Enter an "**Y**" coordinate value <enter>.

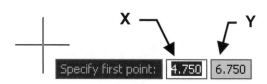

To enter Polar coordinates (from the last point entered)

1. Enter the **distance** value from the last point entered.

2. Press the **Tab** key.

3. Move the cursor in the approximate direction and enter the **angle** value <enter>

Note: Enter an angle value of <u>0-180</u> only and the angle is always <u>positive</u>. (Refer to example on page 11-9)

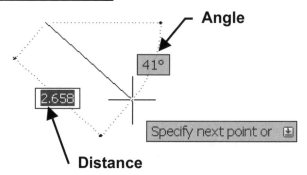

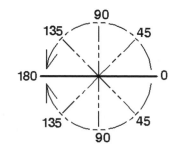

How to specify Absolute or Relative coordinates

<u>To enter absolute coordinates</u> when relative coordinate format is displayed in the tooltip. Enter **#** to temporarily override the setting. (shift key + 3)

<u>To enter relative coordinates</u> when absolute coordinate format is displayed in the tooltip. Enter**@** to temporarily override the setting. (shift key + 2)

<u>To enter absolute world coordinates</u> (WCS), enter ✳

Note about Ortho
You may toggle Ortho On and OFF by holding down the **shift** key.
This is a easy method to use Direct Distance Entry while using Dynamic Input.

USING DYNAMIC INPUT and POLAR COORDINATES

The following is a simple drawing to practice Dynamic Input and Polar coordinates.
Think about how this differs from the basic polar input on page 11-2 and 11-3.

1 Set the Status Bar as follows:
   DYN = ON   All others = OFF

2. Select the **Line** command:

3. Start the Line anywhere in the drawing area.

Line A   4. Move the cursor to the right.
             5. Type 2 <tab> 0 <enter>

Line B   6. Move the cursor up to the right
             7. Type 3 <tab> 45 <enter>

Line C   8. Move the cursor up.
             9. Type 2 <tab>   90<enter>

Line D   10. Move the cursor down to the left.
             11. Type 4 <tab>   135<enter>   (Note: 180 – 45 = 135)

Line E   12. Move the cursor to the left.
             13. Type 1.293 <tab> 180 <enter>

Line F   14. Move the cursor down.
             15. Type 1.293 <tab> 90 <enter>

If you move the cursor around you will notice that the angle value display never exceeds 180.

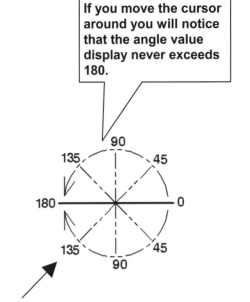

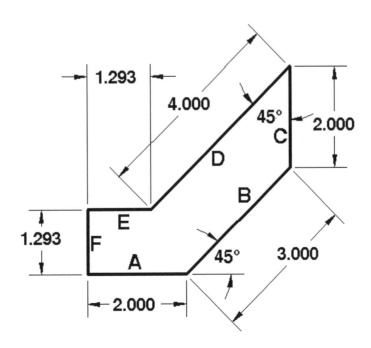

POLAR TRACKING

Polar Tracking can be used instead of **Dynamic Input**. When *Polar Tracking* is "**ON**", a dotted *"tracking"* line and a *"tool tip"* box appear. The tracking line.... "snaps" to a **preset angle increment** when the cursor approaches one of the preset angles. The word *"Polar"* , followed by the *"distance"* and *"angle"* from the last point appears in the box.

(A step by step example is described on the next page.)

SETTING THE ANGLE INCREMENT

1. Right Click on the POLAR button on the Status Bar and select "**SETTINGS**", or select **Tools / Drafting Settings / Polar Tracking** tab. The following dialog box will appear:

2. Set the Increment Angle to: 15.

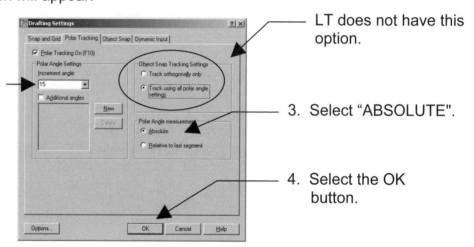

Tracking line

Tool Tip

LT does not have this option.

3. Select "ABSOLUTE".

4. Select the OK button.

POLAR ANGLE SETTINGS

Increment Angle — Choose from a list of Angle increments including 90, 45, 30, 22.5, 18, 15,10 and 5. You will be able to "snap" to multiples of that angle.

Additional Angles — Check this box if you would like to use an angle other than one in the Incremental Angle list. For example: 12.5.

New — You may add an angle by selecting the "New" button. You will be able to snap to this new angle in addition to the incremental Angle selected. But you will not be able to snap to it's multiple. For example, if you selected 7, you would not be able to snap to 14.

Delete — Deletes an Additional Angle. Select the Additional angle to be deleted and then the Delete button.

POLAR ANGLE MEASUREMENT

ABSOLUTE Polar tracking angles are relative to the UCS.

RELATIVE TO LAST SEGMENT Polar tracking angles are relative to the last segment.

USING POLAR TRACKING AND DIRECT DISTANCE ENTRY

1. Use the Polar Tracking Settings from the previous page.
2. Turn all the Status Bar buttons to Off except POLAR. (Note: DYN must be OFF)

SNAP GRID ORTHO POLAR OSNAP OTRACK DUCS DYN LWT MODEL

Select the Line command:

P1 1. Start the Line anywhere in the drawing area.

P2 1. Move the cursor in the direction of P2 until the Tool Tip box displays 30 degrees. (Length is not important yet.)
2. Type 2 <enter> (for the length).

P3 1. Move the cursor in the direction of P3 until the Tool Tip box displays 90 degrees. (Length is not important yet.)
2. Type 2 <enter> (for the length).

P4 1. Move the cursor in the direction of P4 until the Tool Tip box displays 0 degrees. (Length is not important yet.)
2. Type 2 <enter> (for the length).

P5 1. Move the cursor in the direction of P5 until the Tool Tip box displays 150 degrees. (Length is not important yet.)
2. Type 2 <enter> (for the length).

P6 1. Move the cursor in the direction of P6 until the Tool Tip box displays 180 degrees. (Length is not important yet.)
2. Type 2 <enter> (for the length).
3. Then type C for close.

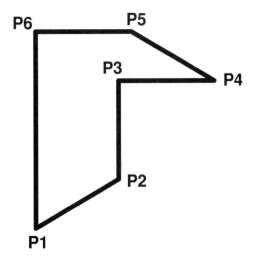

POLAR TRACKING ON or OFF

You may toggle Polar Tracking On or Off using one of the following:
- Left click on the POLAR button on the Status Bar
- Press F10

POLAR SNAP

Polar Snap is used with Polar Tracking to make the cursor snap to specific ***distances*** and ***angles***. If you set <u>Polar Snap distance to 1</u> and <u>Polar Tracking to angle 30</u> you can draw lines 1, 2, 3, 4 units… long at an angle of 30, 60, 90 etc. without typing anything on the command line. You just move the cursor and watch the tool tips.
(A step by step example is described on the next page)

SETTING THE ANGLE INCREMENT

1. Right Click on the POLAR button on the Status Bar and select "SETTINGS" or select **Tools / Drafting Settings / Polar Tracking** tab.
   The following dialog box will appear:

2. Select

3. Set the Increment Angle to: 15

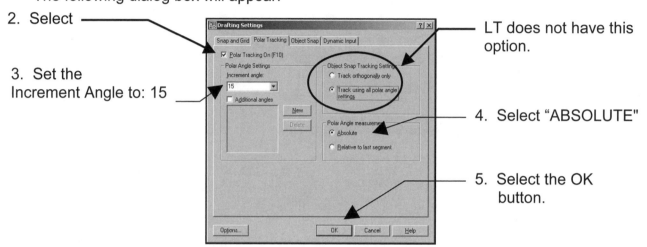

LT does not have this option.

4. Select "ABSOLUTE"

5. Select the OK button.

SETTING THE POLAR SNAP

1. Right Click on the SNAP button on the Status Bar and select "SETTINGS" or select **Tools / Drafting Settings / Snap and Grid** tab.
   The following dialog box will appear:

2. Select

4. Set the Polar Spacing distance

3. Select Polar Snap

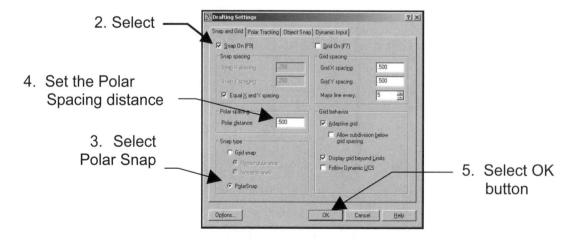

5. Select OK button

SNAP AND GRID - Sets standard snap and grid information
POLAR SPACING - Increment Snap distance when Polar Snap is ON.
STYLE - This setting will be taught in the Advanced course.
TYPE - Sets the Snap to Polar or Grid

USING POLAR TRACKING AND POLAR SNAP

Now let's draw the objects below again, but this time with "Polar Snap" instead of Direct Distance Entry.

1. Use the Polar Tracking and Polar Snap Settings from the previous page.

2. Turn all the Status Bar buttons Off except **SNAP**, **GRID** and **POLAR**.

Select the Line command:

P1
  1. Start the Line anywhere in the drawing area.

P2
  1. Move the cursor in the direction of P2 until the Tool Tip box displays **Polar 2.00 <30°**

P3
  1. Move the cursor in the direction of P3 until the Tool Tip box displays **Polar 2.00 <90°**

P4
  1. Move the cursor in the direction of P4 until the Tool Tip box displays **Polar 2.00 <0°**

P5
  1. Move the cursor in the direction of P5 until the Tool Tip box displays **Polar 2.00 <150°**

P6
  1. Move the cursor in the direction of P6 until the Tool Tip box displays **Polar 2.00 <180°**
  2. Then type C for close.

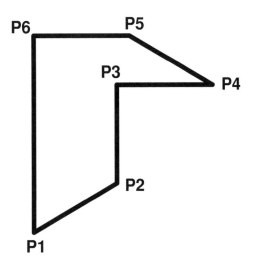

NOTE: You may OVERRIDE the Polar Settings at any time by typing: Polar coordinates (@ Length< Angle) on the Command line.

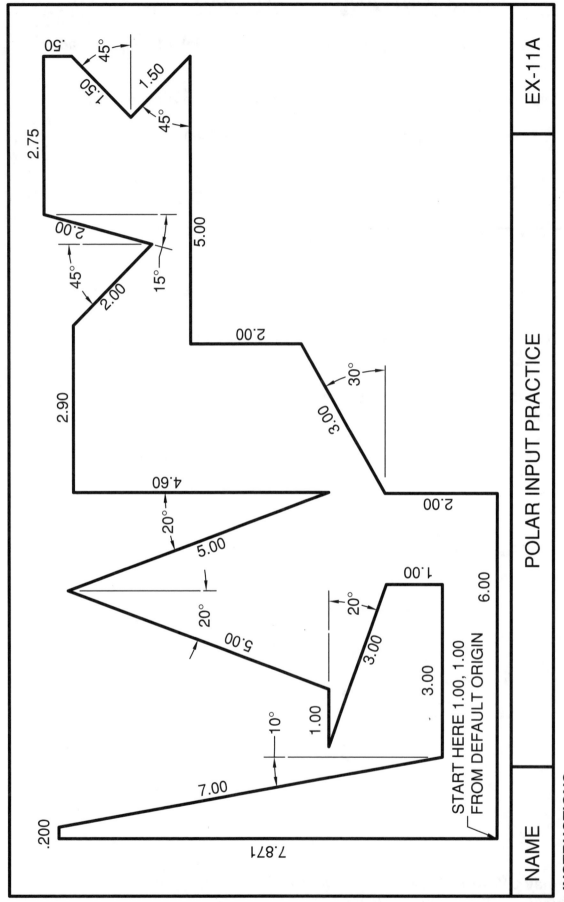

EXERCISE 11A

	POLAR INPUT PRACTICE	EX-11A
NAME		

INSTRUCTIONS:
1. Draw the lines above. Input method is your choice. (Review pages 11-3 and 11-13)
2. Start the first endpoint using ABSOLUTE COORDINATES (1, 1).
3. Save as EX11A and Plot.

11-14

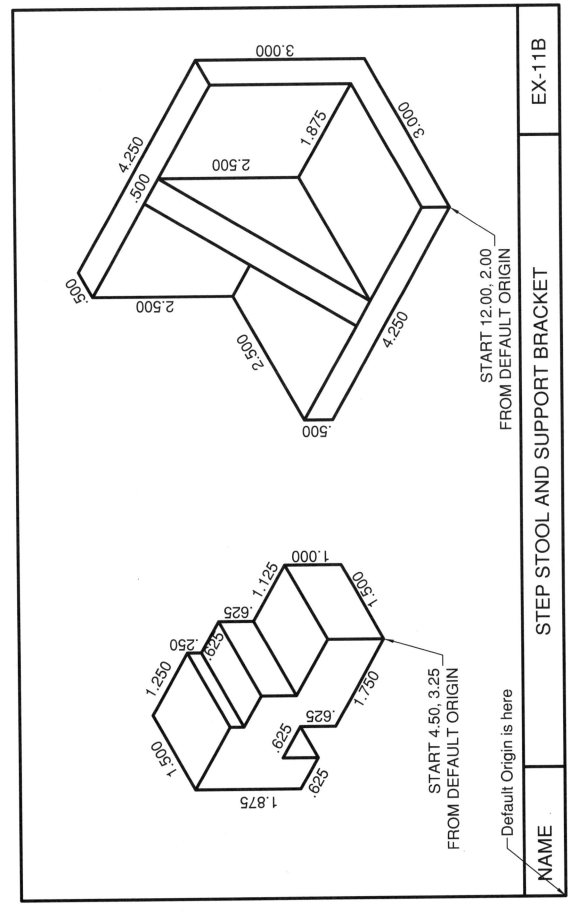

EXERCISE 11B

START 12.00, 2.00
FROM DEFAULT ORIGIN

STEP STOOL AND SUPPORT BRACKET

EX-11B

START 4.50, 3.25
FROM DEFAULT ORIGIN

Default Origin is here

NAME

INSTRUCTIONS:

1. Draw the lines above. Input method is your choice.
2. The Isometric lines are on a 30° angle. (30, 150, 210, 330)
3. Location and size must be accurate.
4. Save as EX 11B and Plot.

11-15

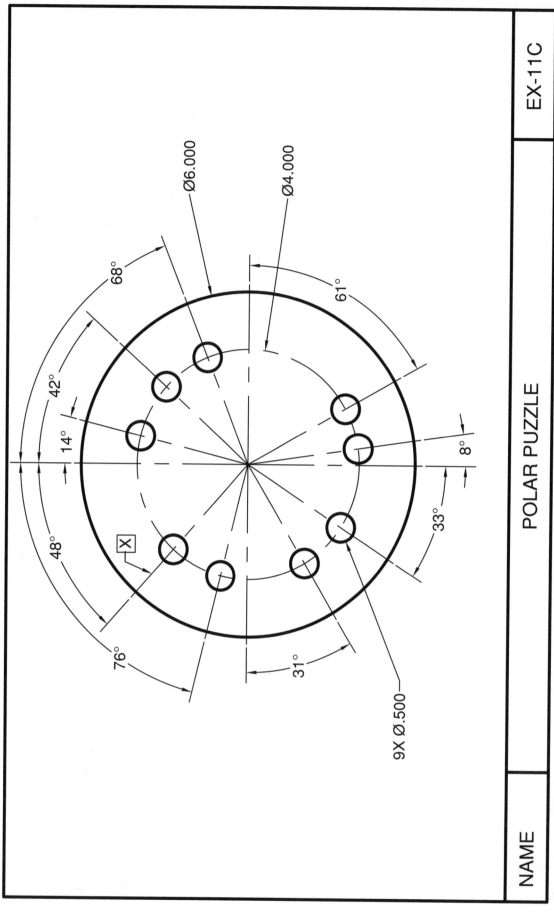

Ø6.000

Ø4.000

68°

61°

42°

14°

8°

48°

33°

X

76°

31°

9X Ø.500

| NAME | POLAR PUZZLE | EX-11C |

EXERCISE 11C

INSTRUCTIONS:

This may be a challenge.

For example, the line marked "X" can be drawn as follows:

1. Place the first endpoint by snapping to the "center" of the Circle.

2. Now type @3.125<138 <enter>.

The length is .125 beyond the radius of the circle and the angle is 90 + 48 =138

Now see if you can figure out the remaining angles. (Refer to 11-3 and 11-13 for angles)

Save as EX 11C and Plot.

LEARNING OBJECTIVES

After completing this lesson, you will be able to:

1. Duplicate an object at a specified distance away.
2. Make changes to an object's properties.
3. Create tables using Single Line and Multiline text.

Template Srg
Cad 2
2008 Spring
Template

LESSON 12

OFFSET

The **OFFSET** command duplicates an object parallel to the original object at a specified distance. You can offset Lines, Arcs, Circles, Ellipses, 2D Polylines and Splines.

EXAMPLES:

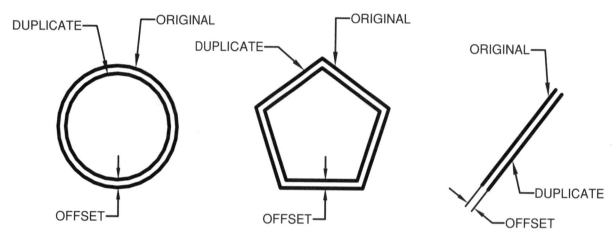

How to use the OFFSET command:

1. Select the OFFSET command using one of the following:

    TYPING = OFFSET
    PULLDOWN = MODIFY / OFFSET
    TOOLBAR = MODIFY

2. Specify offset distance or [Through/Erase/Layer] <Through>: *type the offset distance or select Erase or Layer. (see option on the next page)*

3. Select object to offset or <Exit/Undo>: *select the object to offset.*

4. Specify point on side to offset or [Exit/Multiple/Undo]<Exit>: *Select which side of the original you want the duplicate to appear by placing your cursor and clicking. (See options on the next page)*

5. Select object to offset or [Exit/Undo]<Exit>: *Press <enter> to stop.*

OPTIONS:

Through: Creates an object passing through a specified point.

Erase: Erases the source object after it is offset.

Layer: Determines whether offset objects are created on the <u>current</u> layer or on the layer of the <u>source</u> object. Select <u>Layer</u> and then select <u>current</u> or <u>source.</u>

Multiple: Turns on the multiple offset mode, which allows you to continue creating duplicates of the original without re-selecting the original.

Exit: Exits the Offset command.

Undo: Reverses the previous offset.

OFFSETGAPTYPE (Not available in LT)

When you offset a closed 2D object, such as a rectangle, to create a **larger** object it results in potential gaps between the segments. The **offsetgaptype** system variable controls how these gaps are closed.

To set the offsetgaptype:

Type: ***offsetgaptype <enter>***

Enter one of the following:

0 = Fills the gap by extending the polyline segments. (default setting)

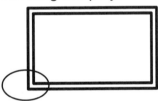

1 = Fills the gap with <u>filleted arc segments</u>. The radius of each arc segment is equal to the offset distance.

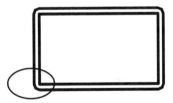

2 = Fills the gap with <u>chamfered line segments.</u> The perpendicular distance to each chamfer is equal to the offset distance.

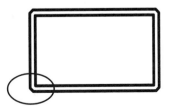

EDITING WITH THE PROPERTIES PALETTE

The **Properties Palette**, shown below, makes it possible to change an object's properties. You simply open the Properties Palette, select an object and you can change any of the properties that are listed.

Open the Properties Palette by double clicking on an object, or use one of the following:

> **TYPING = PROPS or CH**
> **PULLDOWN = MODIFY / PROPERTIES**
> **TOOLBAR = OBJECT PROPERTIES**
> **KEYS = CTRL + 1**

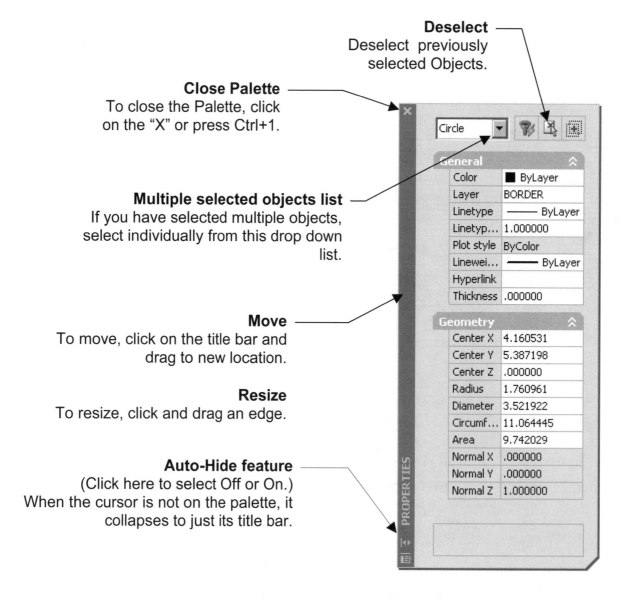

Deselect
Deselect previously selected Objects.

Close Palette
To close the Palette, click on the "X" or press Ctrl+1.

Multiple selected objects list
If you have selected multiple objects, select individually from this drop down list.

Move
To move, click on the title bar and drag to new location.

Resize
To resize, click and drag an edge.

Auto-Hide feature
(Click here to select Off or On.)
When the cursor is not on the palette, it collapses to just its title bar.

Note: The properties listed will depend on the object you have selected.

Examples on the next page.

Example of editing an object using the Properties Palette

1. Draw a 4.00 diameter circle.

2. Double click on the Circle. The Property Palette for the Circle should appear
   *You may change any of the properties listed in the Properties Palette for this object.
   When you press <enter> the circle will change.*

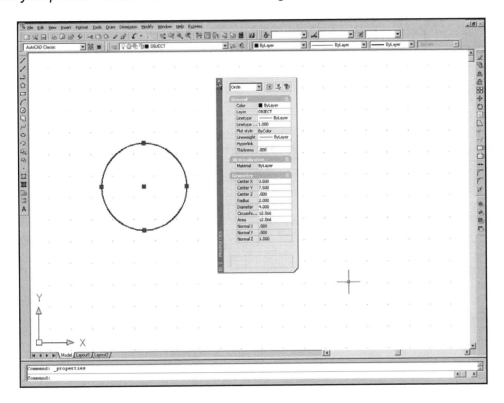

3. Highlight and change the "Radius" to 1.00 and the "Linetype" to hidden <enter>.
   The Circle got smaller and the linetype changed to dashed as shown below.

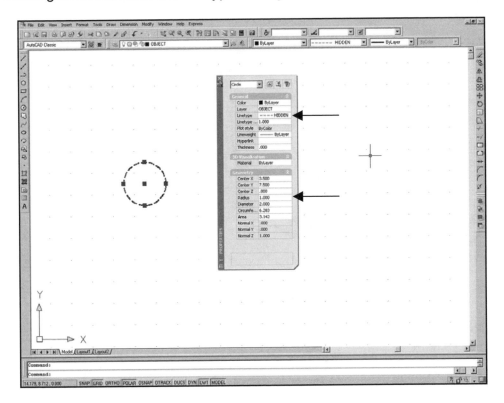

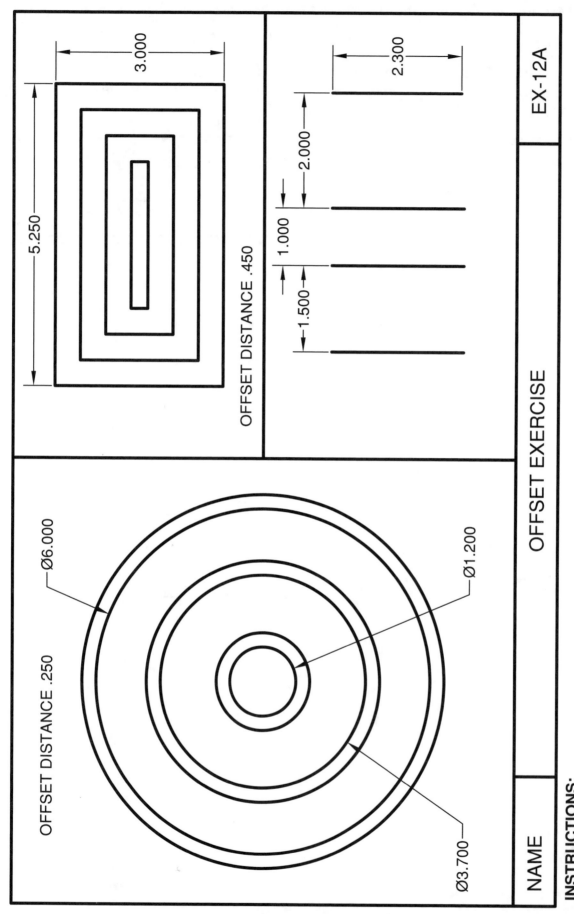

EXERCISE 12A

3.000

5.250

OFFSET DISTANCE .450

2.300

2.000

1.000

1.500

EX-12A

OFFSET EXERCISE

Ø6.000

Ø1.200

OFFSET DISTANCE .250

Ø3.700

NAME

INSTRUCTIONS:
1. Open B-Size.
2. Draw the objects above using:
   Circles, Rectangles, Lines and <u>Offset</u>.
3. Save as: **EX-12A** and Plot.

12-6

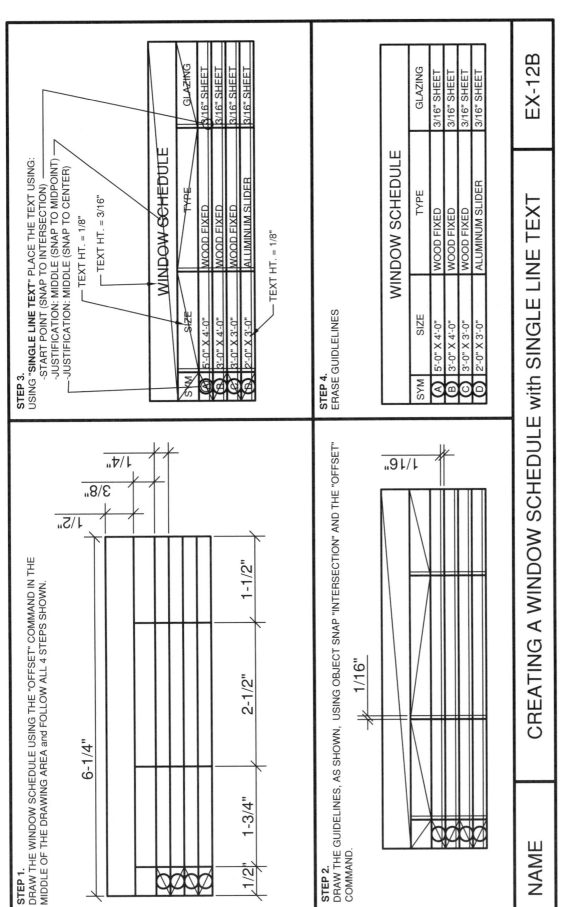

STEP 1.
DRAW THE WINDOW SCHEDULE USING THE "OFFSET" COMMAND IN THE MIDDLE OF THE DRAWING AREA and FOLLOW ALL 4 STEPS SHOWN.

6-1/4"
1/2"
3/8"
1/4"
1-3/4"
2-1/2"
1-1/2"

STEP 2.
DRAW THE GUIDELINES, AS SHOWN, USING OBJECT SNAP "INTERSECTION" AND THE "OFFSET" COMMAND.

1/16"

STEP 3.
USING "**SINGLE LINE TEXT**" PLACE THE TEXT USING:
-START POINT (SNAP TO INTERSECTION)
-JUSTIFICATION: MIDDLE (SNAP TO MIDPOINT)
-JUSTIFICATION: MIDDLE (SNAP TO CENTER)

TEXT HT. = 1/8"
TEXT HT. = 3/16"
TEXT HT. = 1/8"

WINDOW SCHEDULE

SYM	SIZE	TYPE	GLAZING
A	5'-0" X 4'-0"	WOOD FIXED	3/16" SHEET
B	3'-0" X 4'-0"	WOOD FIXED	3/16" SHEET
C	3'-0" X 3'-0"	WOOD FIXED	3/16" SHEET
D	2'-0" X 3'-0"	ALUMINUM SLIDER	3/16" SHEET

STEP 4.
ERASE GUIDELINES

WINDOW SCHEDULE

SYM	SIZE	TYPE	GLAZING
A	5'-0" X 4'-0"	WOOD FIXED	3/16" SHEET
B	3'-0" X 4'-0"	WOOD FIXED	3/16" SHEET
C	3'-0" X 3'-0"	WOOD FIXED	3/16" SHEET
D	2'-0" X 3'-0"	ALUMINUM SLIDER	3/16" SHEET

EX-12B

EXERCISE 12B

NAME	CREATING A WINDOW SCHEDULE with SINGLE LINE TEXT	EX-12B

INSTRUCTIONS:

1. Open B-Size.
2. Follow the steps to draw the Window Schedule above.
   (**Use** Single line text. **Do not** use Multiline text)
3. Use Layer: Border for lines. Layer: Text-Hvy for Headings.
   Layer: Text-Lit for data.
4. Save as: **EX-12B** and **Plot**

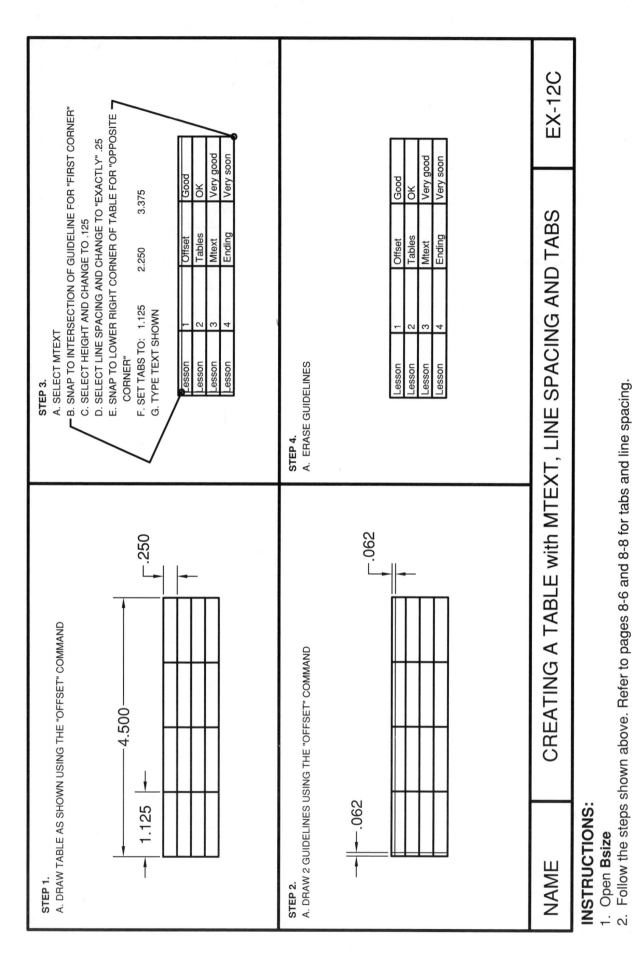

STEP 1.
A. DRAW TABLE AS SHOWN USING THE "OFFSET" COMMAND

4.500
1.125
.250

STEP 2.
A. DRAW 2 GUIDELINES USING THE "OFFSET" COMMAND

.062
.062

STEP 3.
A. SELECT MTEXT
B. SNAP TO INTERSECTION OF GUIDELINE FOR "FIRST CORNER"
C. SELECT HEIGHT AND CHANGE TO .125
D. SELECT LINE SPACING AND CHANGE TO "EXACTLY" .25
E. SNAP TO LOWER RIGHT CORNER OF TABLE FOR "OPPOSITE CORNER"
F. SET TABS TO: 1.125   2.250   3.375
G. TYPE TEXT SHOWN

Lesson	1	Offset	Good
Lesson	2	Tables	OK
Lesson	3	Mtext	Very good
Lesson	4	Ending	Very soon

STEP 4.
A. ERASE GUIDELINES

Lesson	1	Offset	Good
Lesson	2	Tables	OK
Lesson	3	Mtext	Very good
Lesson	4	Ending	Very soon

| NAME | CREATING A TABLE with MTEXT, LINE SPACING AND TABS | EX-12C |

EXERCISE 12C

INSTRUCTIONS:
1. Open **Bsize**
2. Follow the steps shown above. Refer to pages 8-6 and 8-8 for tabs and line spacing.
3. Save as **EX12C** and Plot.

Note: You will learn another method for creating "Tables"
in the Advanced Workbook.

EXERCISE 12D

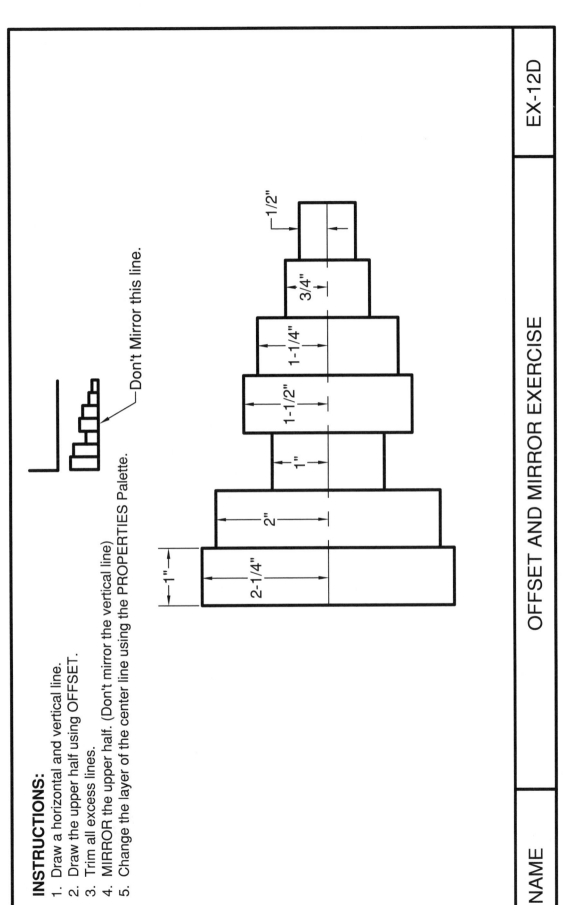

INSTRUCTIONS:
1. Draw a horizontal and vertical line.
2. Draw the upper half using OFFSET.
3. Trim all excess lines.
4. MIRROR the upper half. (Don't mirror the vertical line)
5. Change the layer of the center line using the PROPERTIES Palette.

—Don't Mirror this line.

1/2"

3/4"

1-1/4"

1-1/2"

1"

2"

1"

2-1/4"

| NAME | OFFSET AND MIRROR EXERCISE | EX-12D |

INSTRUCTIONS:
1. Open B-Size.
2. Follow the instructions shown above.
3. Save as: EX-12D and Plot.

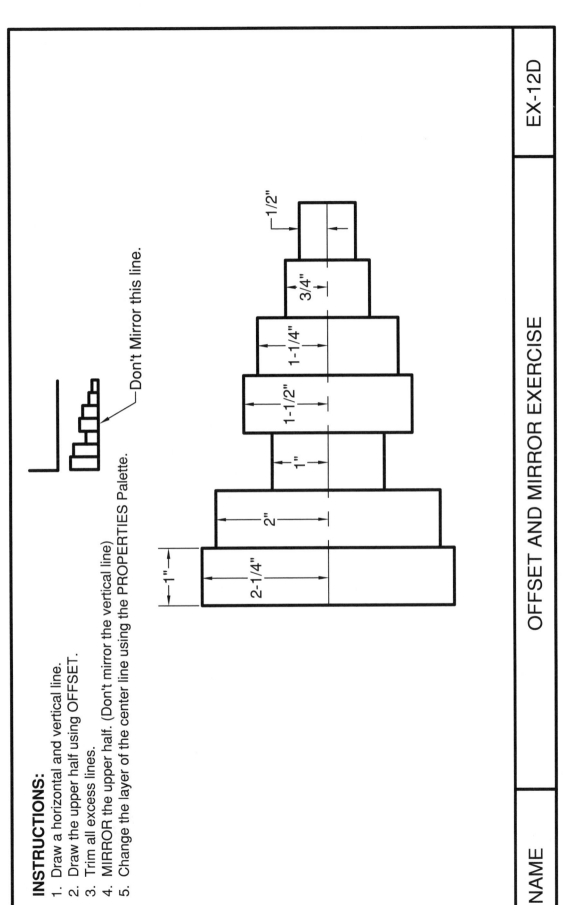

12-9

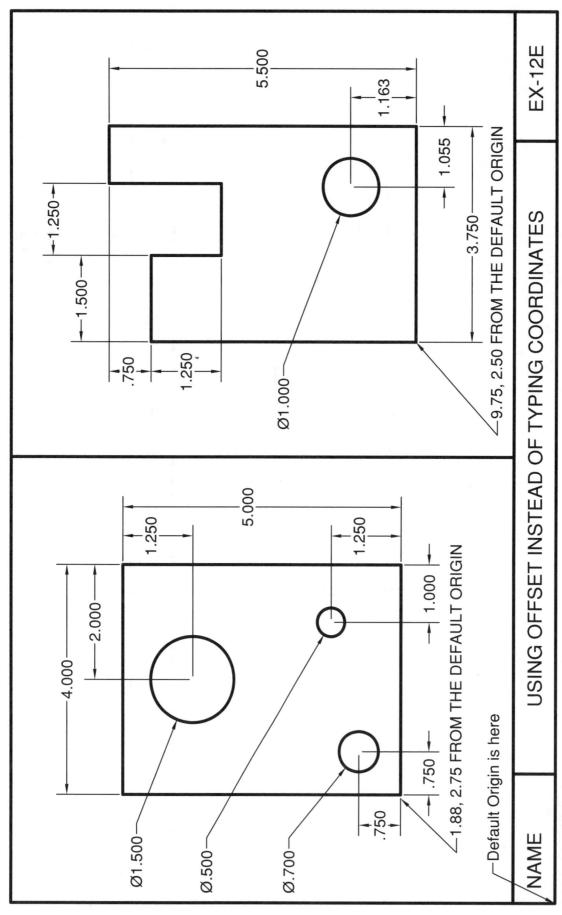

EXERCISE 12E

EX-12E

USING OFFSET INSTEAD OF TYPING COORDINATES

NAME

INSTRUCTIONS:

1. Open B-Size.
2. Draw the Objects above using:
   Offset, Line, Circle and Trim.
   (Do not type coordinates. Create Intersections by offsetting the lines)
3. Save as: **EX-12E** and **Plot.**

Top view labels:

5.500
1.163
1.055
3.750
9.75, 2.50 FROM THE DEFAULT ORIGIN
1.250
1.500
.750
1.250
Ø1.000

Bottom view labels:

5.000
1.250
1.250
2.000
4.000
1.000
.750
.750
1.88, 2.75 FROM THE DEFAULT ORIGIN
Default Origin is here
Ø1.500
Ø.500
Ø.700

12-10

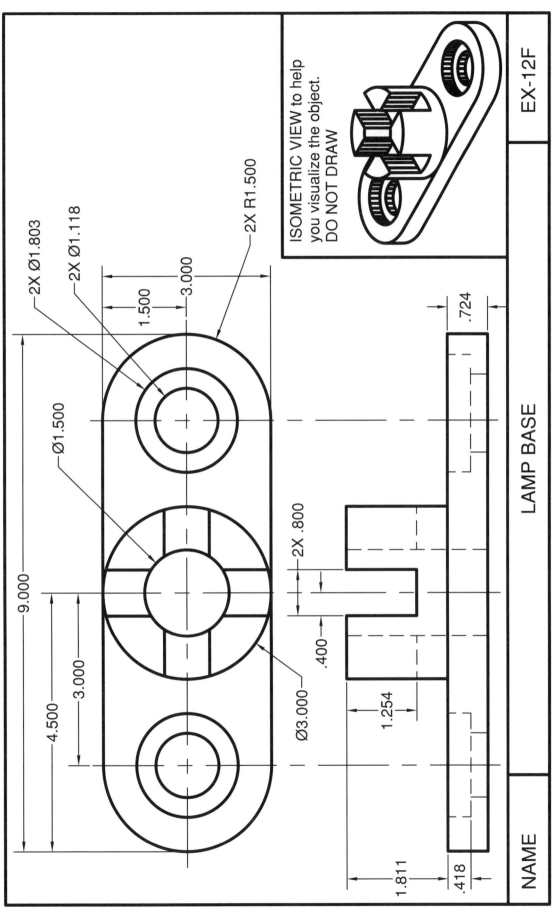

ISOMETRIC VIEW to help
you visualize the object.
DO NOT DRAW

2X Ø1.803

2X Ø1.118

2X R1.500

1.500

3.000

Ø1.500

9.000

4.500

3.000

Ø3.000

2X .800

.400

1.254

1.811

.418

.724

EX-12F

EXERCISE 12F

NAME

LAMP BASE

INSTRUCTIONS:

1. Open B-Size.

2. Draw the Objects above.
   Practice the "Layer" option by making the "Hidden" or "Center" layer current when
   offsetting. (See 12-3) Or use the Properties Palette to modify the linetype.
   You decide the best method.   Note: It is Symmetrical.

3. Save as: **EX-12F** and **Plot.**

NOTES:

LEARNING OBJECTIVES

After completing this lesson, you will be able to:

1. Create multiple copies in a Rectangular or Circular pattern.

LESSON 13

ARRAY

The ARRAY command allows you to make multiple copies in a **RECTANGULAR** or Circular **(POLAR)** pattern. The maximum limit of copies per array is 100,000. This limit can be changed but should accommodate most users.

RECTANGULAR ARRAY

The method allows you to make multiple copies of object(s) in a rectangular pattern. You specify the number of rows (horizontal), columns (vertical) and the offset distance between the rows and columns. The offset distances will be equally spaced.

Offset Distance is sometimes tricky to understand. *Read this carefully*. The offset distance is the distance from a specific location on the original to that same location on the invisible copy. It is not just the space in between the two. Refer to the example below.

To use the rectangular array command you will select the object(s), specify how many rows and columns desired and the offset distance for the rows and the columns. **Step by step instructions on page 13-3**.

Example of Rectangular Array:

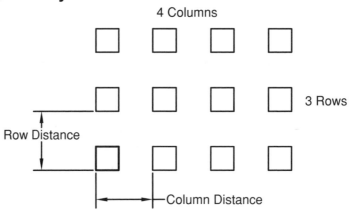

Example of Rectangular Array on an angle:
Notice the copies do not rotate. The Angle is only used to establish the placement.

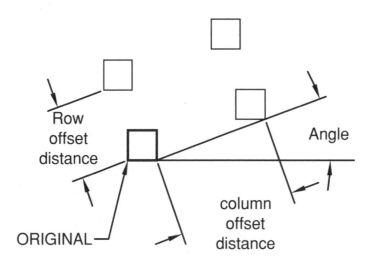

ARRAY (Continued)

RECTANGULAR ARRAY

1. Select the ARRAY command using one of the following:

> **TYPE = ARRAY**
> **PULLDOWN = MODIFY / ARRAY**
> **TOOLBAR = MODIFY**

The dialog box shown below will appear.

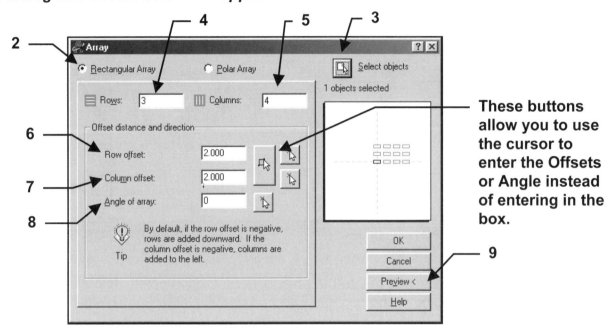

These buttons allow you to use the cursor to enter the Offsets or Angle instead of entering in the box.

2. Select "**Rectangular Array**".

3. Select the **"Select Objects"** button.
   This will take you back to your drawing. Select the objects to Array then <enter>.

4. Enter the number of rows.

5. Enter the number of columns.

6. Enter the row offset. (The distance from a specific location on the original to that same specific location on the future copy.) See example on pg. 13-2.

7. Enter the column offset. (The distance from a specific location on the original to that same specific location on the future copy.) See example on pg. 13-2.

8. Enter an angle if you would like the array to be on an angle.

9. Select the Preview button.
   If it looks correct, select the Accept button. If it is not correct, select the Modify button, make the necessary corrections and preview again.
   (Note: If the Preview button is gray, you have forgotten to "Select objects"- #3)

ARRAY (continued)

POLAR ARRAY

This method allows you to make multiple copies in a circular pattern. You specify the total number of copies to fill a specific Angle or specify the angle between each copy and angle to fill.

To use the polar array command you select the object(s) to copy, specify the center of the array, specify the number of copies or the angle between the copies, the angle to fill and if you would like the copies to rotate as they are copied.

Example of Polar Array

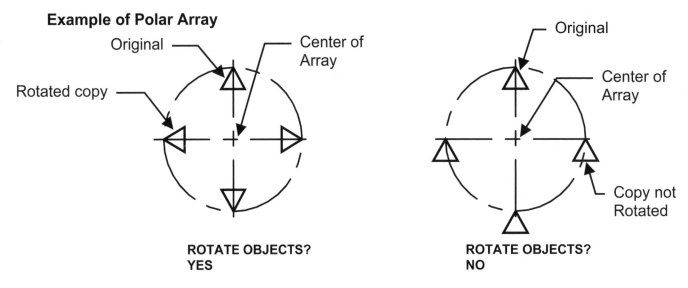

ROTATE OBJECTS?
YES

ROTATE OBJECTS?
NO

Note: the two examples above use the objects default base point. The example below specifies the base point for the copies. The end result is different from the example above on the right.

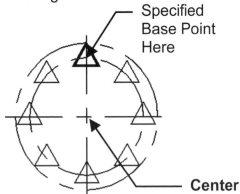

ROTATE OBJECTS?
NO
(Specified base point)

DEFAULT OBJECT BASE POINTS	
Type of Object	**Default Base Pt**.
Arc, Circle, Ellipse	Center
Polygon, Rectangle	First Corner
Line, Polyline, Donut	Starting Point
Text, Block	Insertion Point

Note: If you select multiple objects the base Point of the last object selected is used to contruct the array.

The difference between "Center Point" and "Object Base Point" is sometimes confusing. When specifying the Center Point, try to visualize the copies already there. Now, in your mind, try to visualize the center of that array. In other words, it is the Pivot Point from which the copies will be placed around. The Base Point is different. It is located on the original object.

ARRAY (continued)

POLAR ARRAY

1. Select the ARRAY command using one of the following:

> **TYPE = ARRAY**
> **PULLDOWN = MODIFY / ARRAY**
> **TOOLBAR = MODIFY**

The dialog box shown below will appear.

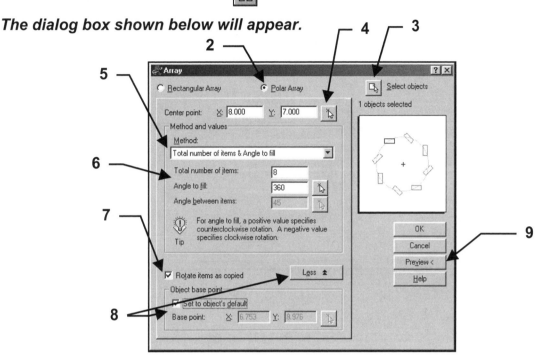

2. Select **" Polar Array".**

3. Select the **"Select Objects"** button.
    This will take you back to your drawing. Select the objects to Array then <enter>.

4. Press the "Center Point" button and select the center point with the cursor or enter
    the X and Y coordinates in the "Center Point" boxes.

5. Select the method.

6. Enter the "Total number of items", "Angle to fill" or "Angle between items".

7. Select whether you want the items rotated as copied or not. (See pg. 13-4 for
    explanation.)

8. Accept the object's default base point or enter the X and Y coordinates.
    (Select the "More" button to show this area. Select the "Less" button to not show.)

9. Select the Preview button.
    If it looks correct, select the Accept button. If it is not correct, select the Modify
    button, make the necessary corrections and preview again.
    (Note: If the Preview button is gray, you have forgotten to select objects #3)

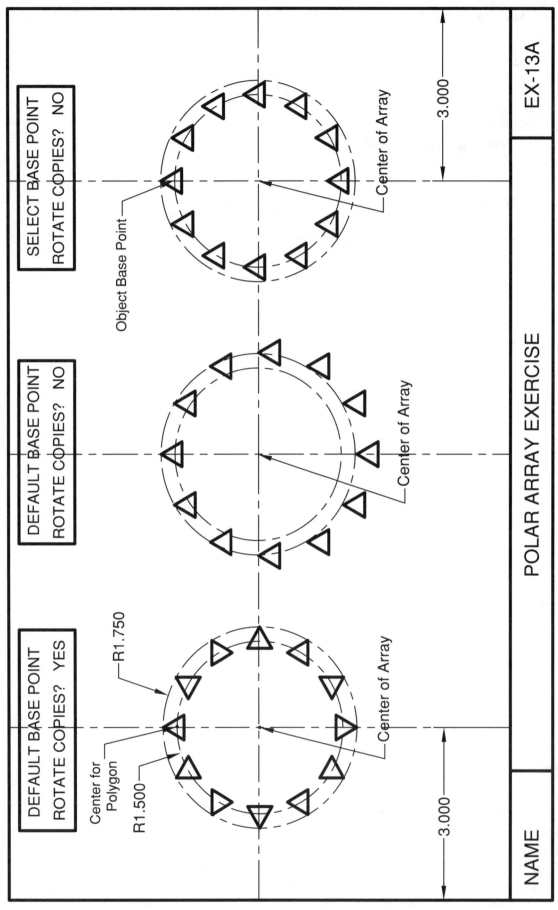

DEFAULT BASE POINT
ROTATE COPIES? YES

DEFAULT BASE POINT
ROTATE COPIES? NO

SELECT BASE POINT
ROTATE COPIES? NO

Center for Polygon

R1.750

R1.500

Center of Array

Center of Array

Center of Array

Object Base Point

3.000

3.000

POLAR ARRAY EXERCISE

NAME

EX-13A

EXERCISE 13A

INSTRUCTIONS:

1. Open B-Size.
2. Draw the Centerlines and Circles first. (use layer: Center)
3. Draw the **Original Polygon** (Inscribed, R.1/4") at the **12:00 position**.
4. Using Polar Array, array the original Polygon. Number of items: **12**
5. Save as: **EX-13A** and Plot

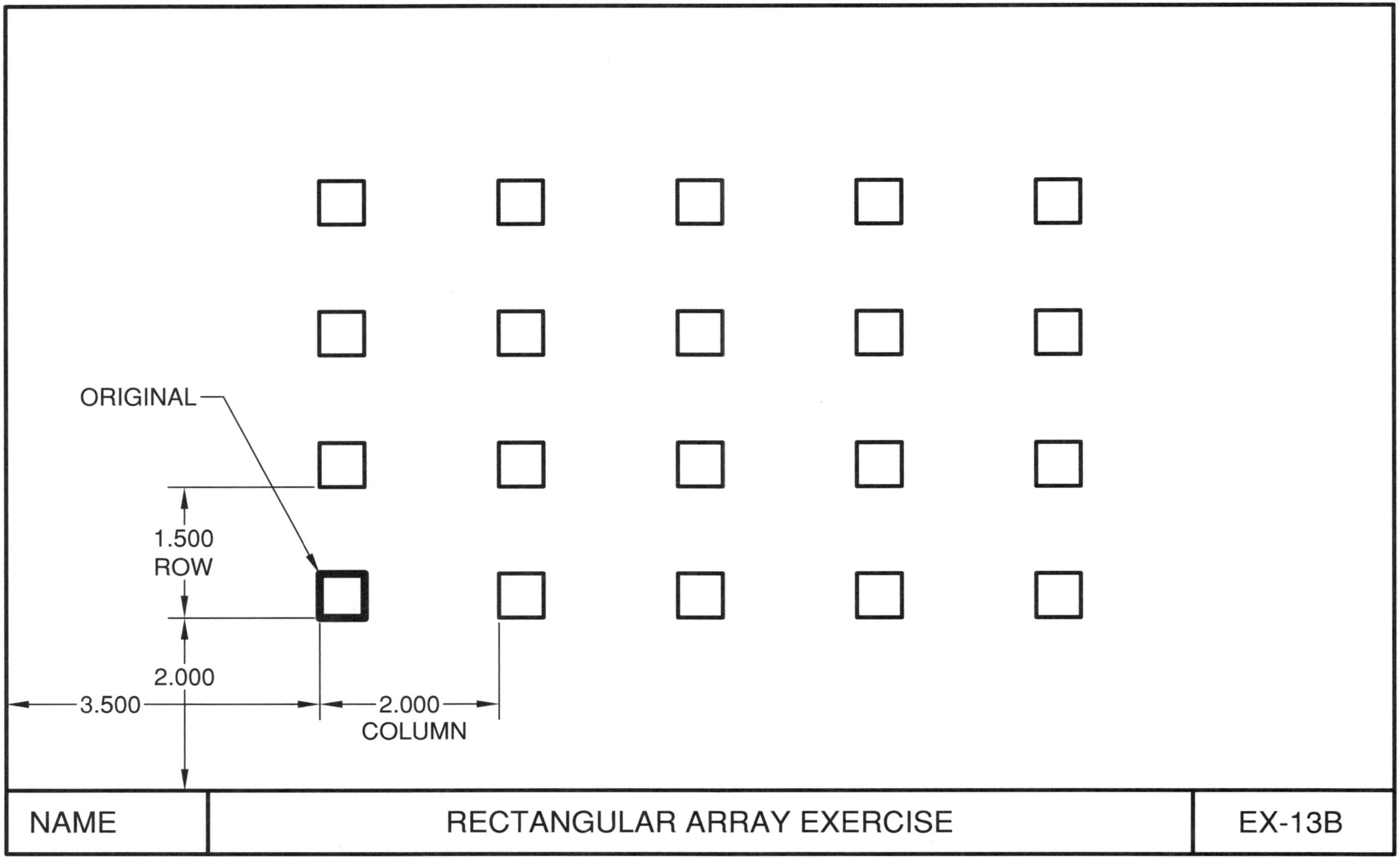

EXERCISE 13B

EX-13B
RECTANGULAR ARRAY EXERCISE
NAME

INSTRUCTIONS:

1. Open B-Size.
2. Draw the lower left RECTANGLE first. .500 Square. (layer: Object)
3. Using Rectangular Array, array the original Rectangle as shown.
   No. of Rows = 4   No. of Columns = 5
   Distance between Rows = 1.500   Distance between columns = 2.000
4. Save as: **EX-13B** and Plot.

ORIGINAL

1.500
ROW

2.000
COLUMN

2.000

3.500

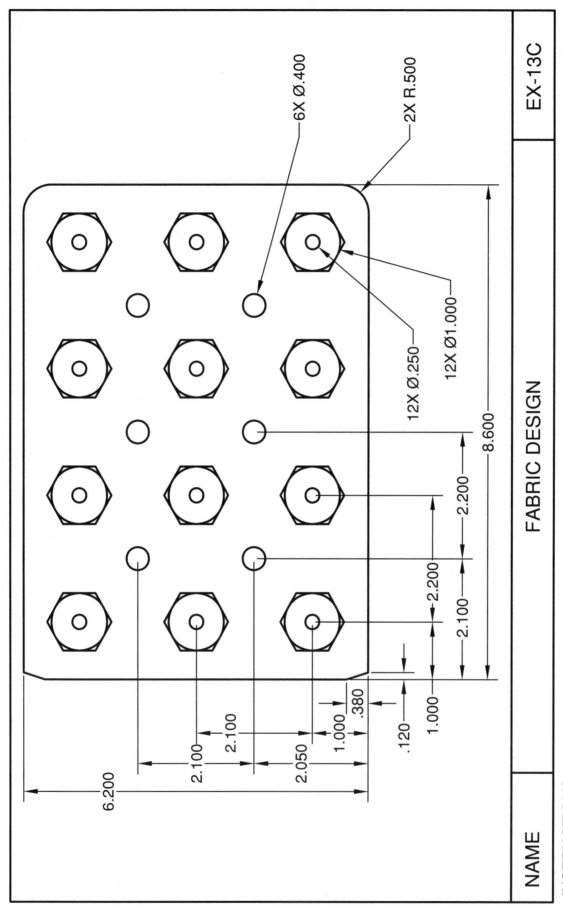

EXERCISE 13C

EX-13C

FABRIC DESIGN

6X Ø.400

2X R.500

12X Ø.250

12X Ø1.000

8.600

2.200

2.200

2.100

1.000

.120

.380

1.000

2.050

2.100

2.100

6.200

NAME

INSTRUCTIONS:

1. Open B-Size.
2. Draw the objects above using the most efficient methods.
3. Save as: **EX-13C** and plot.

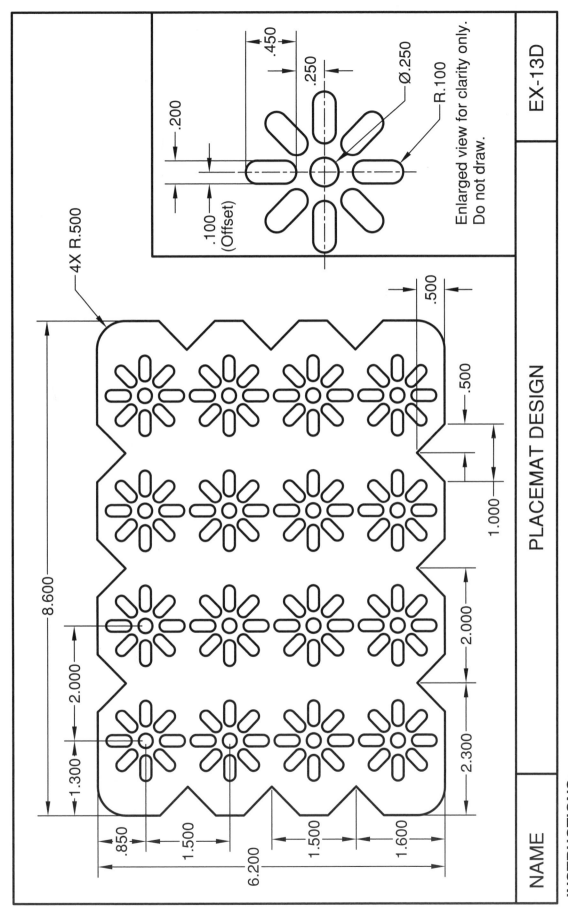

EXERCISE 13D

.450

.200

.250

Ø.250

R.100

.100
(Offset)

Enlarged view for clarity only.
Do not draw.

EX-13D

4X R.500

.500

.500

1.000

2.000

2.300

8.600

2.000

1.300

.850

1.500

1.500

1.600

6.200

PLACEMAT DESIGN

NAME

INSTRUCTIONS:

1. Open B-Size.
2. Draw the objects above using the most efficient methods.
3. Save as: **EX-13D** and plot.

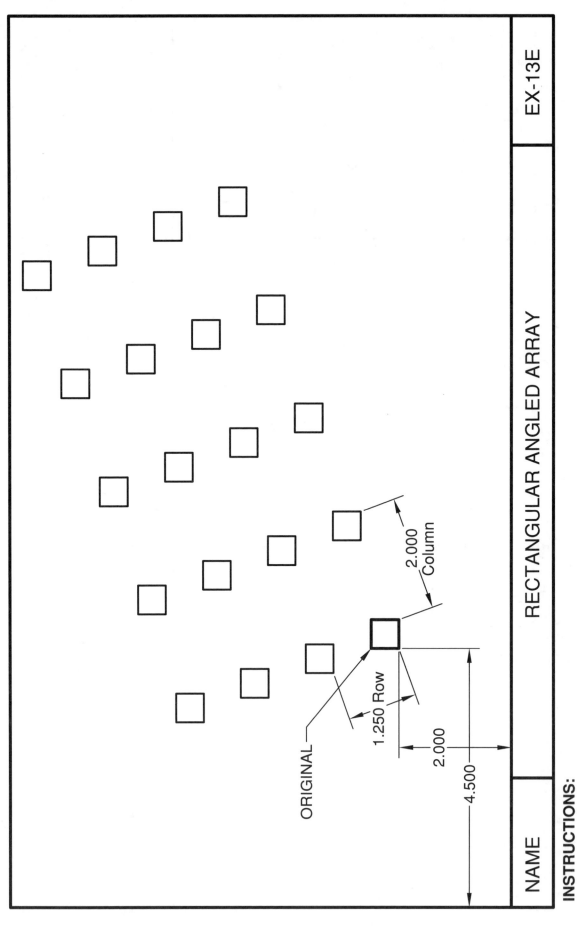

RECTANGULAR ANGLED ARRAY

EX-13E

EXERCISE 13E

NAME

INSTRUCTIONS:

1. Open B-Size.
2. Draw the lower left RECTANGLE first. .500 Square. (Use layer: Object)
3. Using Rectangular Array, array the original Rectangle as shown.
   No. of Rows = 4   No. of Columns = 5   Angle = 20
   Distance between Rows = 1.250   Distance between columns = 2.000
4. Save as: **EX-13E** and plot.

LEARNING OBJECTIVES

After completing this lesson, you will be able to:

1. Make an existing object larger or smaller proportionately.
2. Stretch or compress an existing object.
3. Rotate an existing object to a specific angle.

LESSON 14

SCALE

The **SCALE** command is used to make objects larger or smaller <u>proportionately</u>. You may scale using a scale factor or a reference length. You must also specify a base point. Think of the base point as a stationary point from which the objects scale. It does not move.

1. Select the SCALE command using one of the following:

   TYPE = SCALE
   PULLDOWN = MODIFY / SCALE
   TOOLBAR = MODIFY

SCALE FACTOR
   Command: _scale
2. Select objects: *select the object(s) to be scaled*
3. Select objects: *select more object(s) or <enter> to stop*
4. Specify base point: *select the stationary point on the object*
5. Specify scale factor or [Copy/Reference]: *type the <u>scale factor</u> <enter>*

If the scale factor is greater than 1, the objects will increase in size.
If the scale factor is less than 1, the objects will decrease in size.

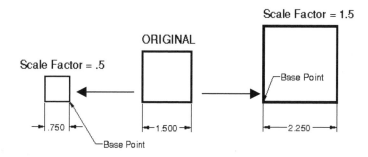

REFERENCE
Command: _scale
2. Select objects: *select the object(s) to be scaled*
3. Select objects: *select more object(s) or <enter> to stop*
4. Specify base point: *select the stationary point on the object*
5. Specify scale factor or [Copy/Reference]: *select Reference*
6. Specify reference length <1>: *specify a <u>reference</u> length*
7. Specify new length: *specify the new length*

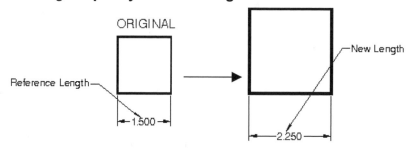

<u>COPY</u> creates a duplicate of the selected object. The duplicate is directly on top of the original. The duplicate will be scaled. The Original remains the same.

STRETCH

The **STRETCH** command allows you to stretch or compress object(s). Unlike the Scale command, you can alter an objects proportions with the Stretch command. In other words, you may increase the length without changing the width and vice versa.

Stretch is a very valuable tool. Take some time to really understand this command. It will save you hours when making corrections to drawings.

When selecting the object(s) you must use a **CROSSING** window.
Objects that are crossed, will *stretch.*
Objects that are totally enclosed, will *move*.

1. Select the STRETCH command using one of the following:

    TYPE = S
    PULLDOWN = MODIFY / STRETCH
    TOOLBAR = MODIFY

    Command: _stretch
2. Select objects to stretch by crossing-window or crossing-polygon...
    Select objects: *select the first corner of the crossing window*
3. Specify opposite corner: *specify the opposite corner of the crossing window*
4. Select objects: *<enter>*
5. Specify base point or [Displacement] <Displacement>: *select a base point (where it stretches from)*
6. Specify second point or <use first point as displacement>:*type coordinates or place location with cursor*

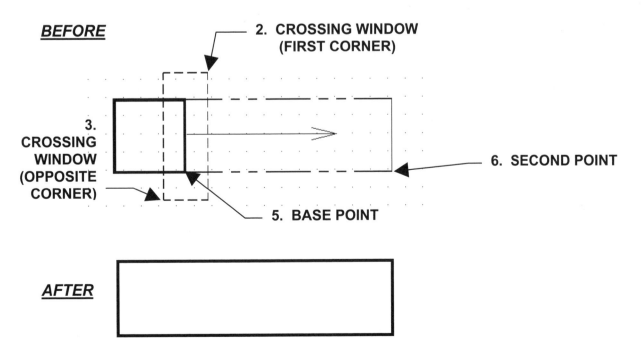

BEFORE

2. **CROSSING WINDOW (FIRST CORNER)**

3. **CROSSING WINDOW (OPPOSITE CORNER)**

5. **BASE POINT**

6. **SECOND POINT**

AFTER

ROTATE

The **ROTATE** command is used to rotate objects around a Base Point. (pivot point)
After selecting the objects and the base point, you will enter the rotation angle or select a reference angle followed by the new angle.
A **Positive** rotation angle revolves the objects **Counter- Clockwise**.
A **Negative** rotation angle revolves the objects **Clockwise**.

Select the ROTATE command using one of the following:

> **TYPE =RO**
> **PULLDOWN = MODIFY / ROTATE**
> **TOOLBAR = MODIFY**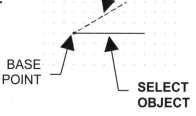

ROTATION ANGLE OPTION
> Command: _rotate
1. Current positive angle in UCS: ANGDIR=counterclockwise ANGBASE=0
   Select objects: *select the object to rotate.*
2. Select objects: *select more object(s) or <enter> to stop.*
3. Specify base point: *select the base point (pivot point).*
4. Specify rotation angle or [Copy/Reference]<0>:
   > *type the angle of rotation.*

REFERENCE OPTION
> Command: _rotate
1. Current positive angle in UCS: ANGDIR=counterclockwise ANGBASE=0
   Select objects: *select the object to rotate.*
2. Select objects: *select more object(s) or <enter> to stop.*
3. Specify base point: *select the base point (pivot point).*
4. Specify rotation angle or [Reference]: *select Reference.*
5. Specify the reference angle <0>: *Snap to the reference object (1) and (2).*
6. Specify the new angle: *drag the object and snap to the new angle.*

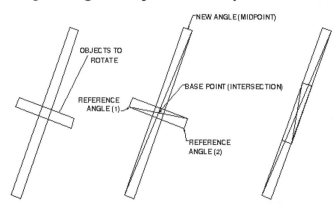

COPY OPTION
Creates a duplicate of the selected object(s). The duplicate is directly on top of the original. The duplicate will be rotated. The Original remains the same.

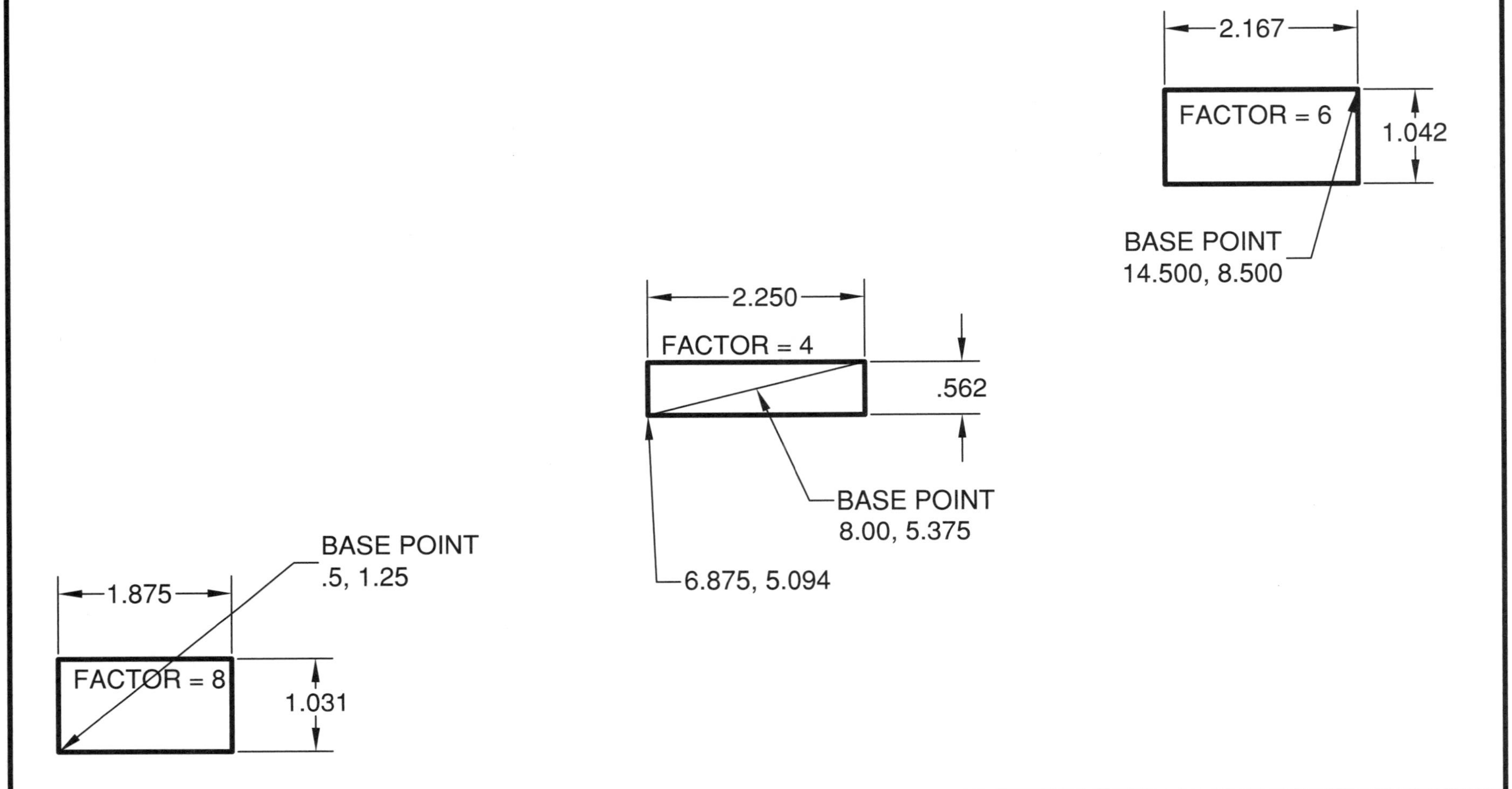

EXERCISE 14A

2.167

1.042

FACTOR = 6

BASE POINT
14.500, 8.500

.562

2.250

FACTOR = 4

BASE POINT
8.00, 5.375

6.875, 5.094

BASE POINT
.5, 1.25

1.031

1.875

FACTOR = 8

EX-14A

SCALE FACTOR EXERCISE

NAME

INSTRUCTIONS:

1. Open B-Size.

2. Draw the 3 Rectangles above. (Size and location is very important)

3. Scale each Rectangle using:

   a. Modify / Scale   b. Use Factors and Base points shown.

4. Save as: **EX-14A** and plot.

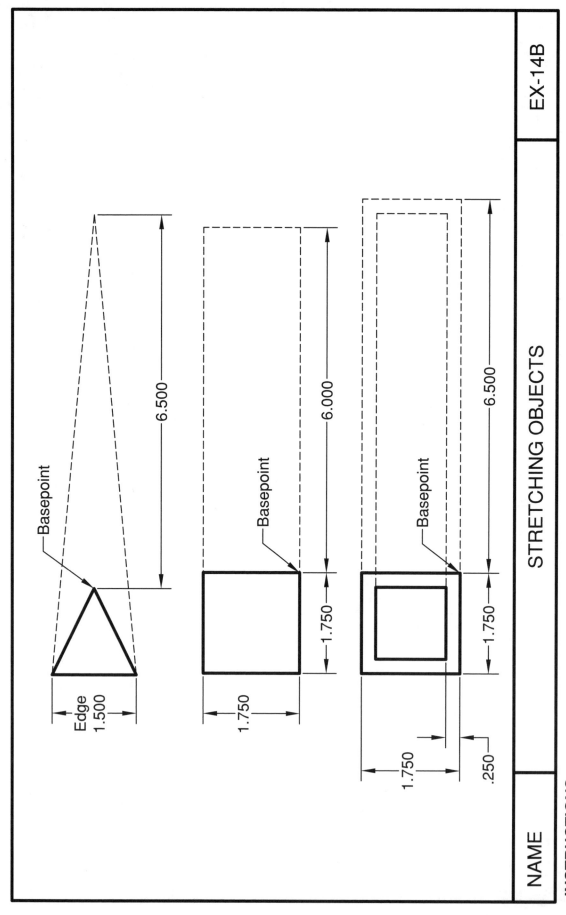

STRETCHING OBJECTS

EXERCISE 14B

EX-14B

NAME

INSTRUCTIONS:

1. Open B-Size.
2. Draw the 3 dark objects on the left, then STRETCH them to the new lengths, represented with dashed lines.
   (Remember to use a CROSSING WINDOW)
3. Save as: **EX-14B** and plot.

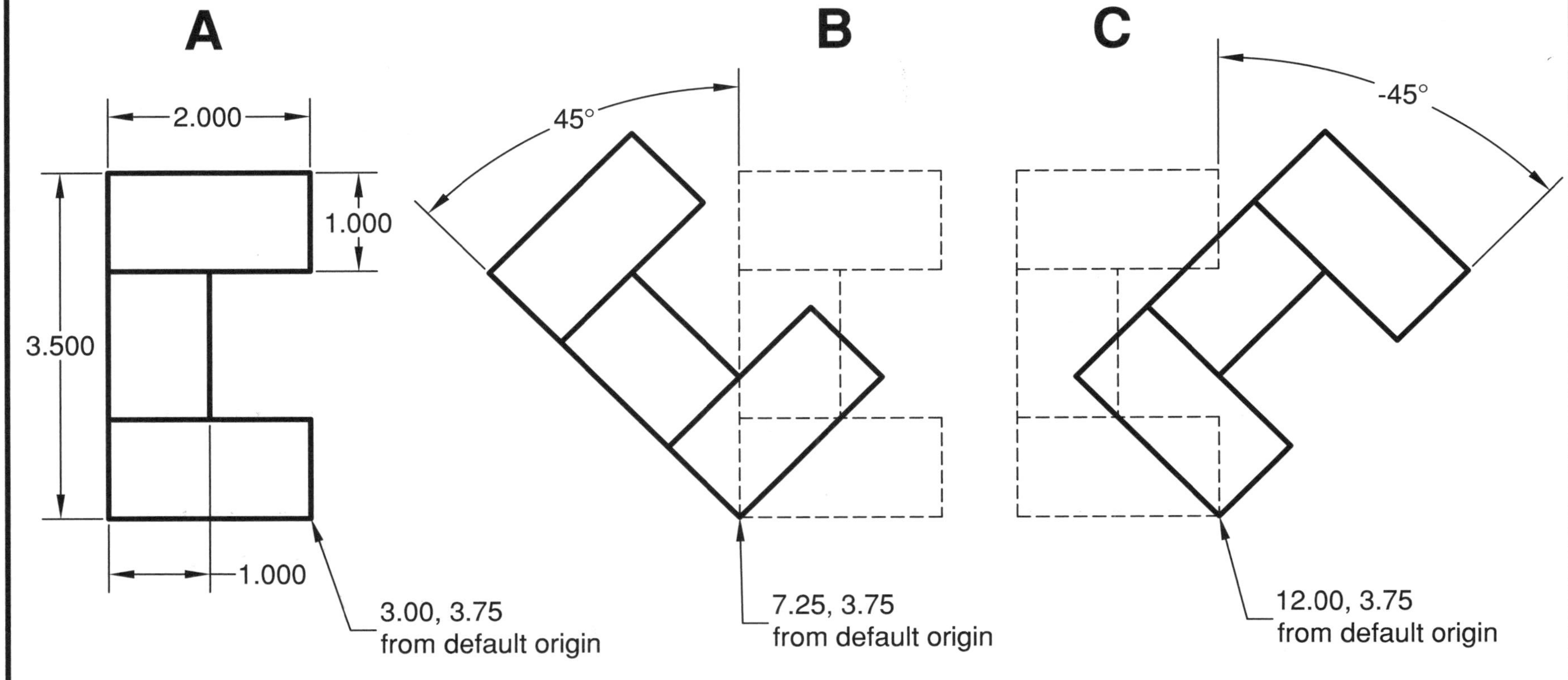

A

1.000

2.000

3.500

1.000

3.00, 3.75
from default origin

B

45°

7.25, 3.75
from default origin

C

-45°

12.00, 3.75
from default origin

ROTATING OBJECTS

EXERCISE 14C

NAME

INSTRUCTIONS:

1. Open B-Size.
2. Draw the object at position A.
3. Copy the first drawing to positions B and C. (Copy location represented with dashed lines)
4. Rotate the copies as shown.
5. Save as: **EX-14C** and plot.

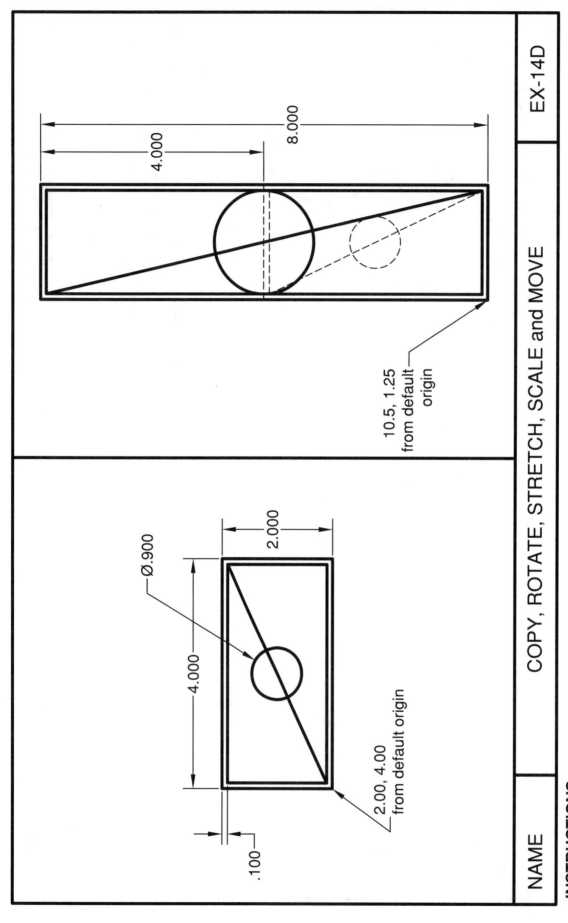

EXERCISE 14D

| NAME | COPY, ROTATE, STRETCH, SCALE and MOVE | EX-14D |

8.000

4.000

10.5, 1.25
from default
origin

Ø.900

2.000

4.000

.100

2.00, 4.00
from default origin

INSTRUCTIONS:

1. Draw the objects on the left. (layer: Object)
2. Copy these objects and move to the right side.
3. Rotate 90° and Stretch.
4. Move the Circle to proper location and Scale. (factor: 2)
5. Save as: **EX-14D** and plot.

14-8

LEARNING OBJECTIVES

After completing this lesson, you will be able to:

1. Cross hatch a section view.
2. Add Gradient filled areas.
3. Solidly fill an area.
4. Make changes to a hatch set already in the drawing.

LESSON 15

HATCH

The **BHATCH** command is used to create hatch lines for section views or filling areas with specific patterns.

To draw **hatch** you must start with a closed boundary. A closed boundary is an area completely enclosed by objects. A rectangle would be a closed boundary. You simply pick inside the closed boundary. BHATCH locates the area and automatically creates a temporary polyline around the outline of the hatch area. After the hatch lines are drawn in the area, the temporary polyline is automatically deleted. (Polylines are discussed in Lesson 23)

Note: A Hatch set is one object. If you explode it, it will return to many objects.

1. Select the BHATCH command using one of the following:

TYPE = BH
PULLDOWN = DRAW / HATCH
TOOLBAR = DRAW

The following dialog box appears:

Gradient tab not available in version "LT"

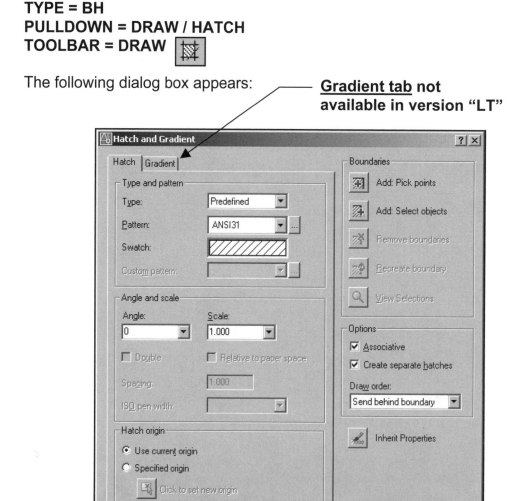

15-2

HATCH (continued)

2. Select the hatch "TYPE"

Select one of the following:
PREDEFINED, USER DEFINED or CUSTOM
(Description of each is listed below.)

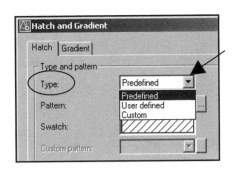

a. *PREDEFINED*

AutoCAD has many predefined hatch patterns. These patterns are stored in the acad.pat and acadiso.pat files. (You may also purchase patterns from other software companies.)

> **Note:** Using Hatch patterns will greatly increase the size of the drawing file. So use them conservatively.

To select a pattern by name, click on the "Pattern" down arrow. A drop-down list of available patterns will appear.

To select a pattern by appearance, click on the (...) button. This will display the "Hatch Pattern Palette" dialog box.

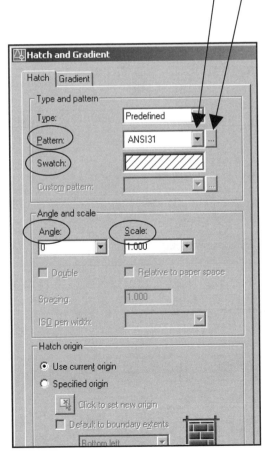

Predefined Pattern Properties

Pattern
This box displays the name of the Pattern you selected.

Swatch
The selected pattern is displayed here.

Angle
This determines the rotation angle of the pattern. A pre-designed pattern has a default angle of 0. If you change this Angle it will rotate the pattern relative to its original design.

Scale
The value in this box is the scale factor. A good starting point would be to enter the scale factor of the drawing. Such as: If the drawing will be plotted at ¼" = 1, "4" is the scale factor. (Drawing Scale factors will be discussed in Lesson 27)

HATCH (continued

b. *USER DEFINED*

This selection allows you to simply draw continuous lines. (No special pattern) You specify the Angle and the Spacing between the lines. (This selection does not increase the size of the drawing file like Predefined)

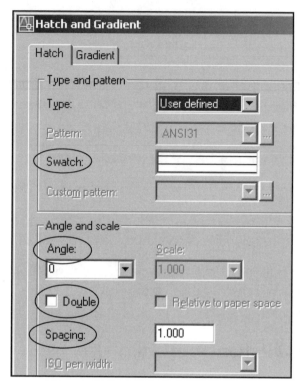

User defined Properties

Swatch
A sample of the angle and spacing settings is displayed here.

Angle
Specify the actual angle of the hatch lines. (0 to 180)

Spacing
Specify the actual distance between each hatch line.

Double
(This option is only available when using "User Defined")
If this box is checked, the hatch set will be drawn first at the angle specified, then a second hatch set will be drawn rotated 90 degrees relative to the first hatch set, creating a criss-cross affect.

3. Select the Options.

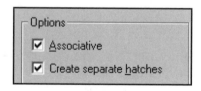

Associative: The hatch set is one entity and if the boundary size is changed the hatch will automatically change to the new boundary shape.

Create separate hatches: Controls whether HATCH creates a single hatch object or separate hatch objects when selecting several closed boundaries.

HATCH (continued)

4. Select the Area you want to Hatch using "Pick Points or Select Objects.

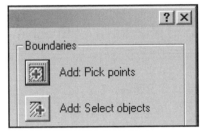

a. **Add: Pick points (You will use this option primarily)**
Select the **Pick points** box then select a point inside the area you want to hatch. A boundary will automatically be determined.

b. **Add: Select objects**
Select the boundary by selecting the object(s).

Click inside the area

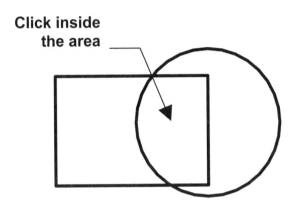

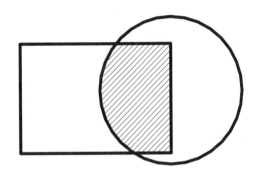

Gap Tolerance

If a warning appears stating "Valid hatch boundary not found" you probably have a gap in the boundary. If this occurs you can find the gap and fix it or you can set the Gap Tolerance to a value from 0 to 5000. (The default value is 0.) Any gaps equal to or smaller than the value you specify are ignored, and the boundary is treated as closed.

To set the gap tolerance: Select the "More Options" button.

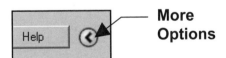 **More Options**

5. Preview the Hatch.

a. After you have selected the boundary, press the right mouse button and select "**Preview**" from the short cut menu. This option allows you to preview the hatch set before it is actually applied to the drawing.

b. If the preview is not what you expected, press the **ESC** key, make the changes and preview again.

c. When you are satisfied, **right click** or press **<enter>** to accept.

Note: It is always a good idea to take the extra time to preview the hatch. It will actually save you time in the long run.

GRADIENT FILLS   (not available in the LT version)

Gradients are fills that gradually change from dark to light or from one color to another. Gradient fills can be used to enhance presentation drawings, giving the appearance of light reflecting on an object, or creating interesting backgrounds for illustrations.

Gradients are definitely fun to experiment with but you will have to practice a lot to achieve complete control.

1. Select the BHATCH command using one of the following:

   TYPE = BH
   PULLDOWN = DRAW / HATCH
   TOOLBAR = DRAW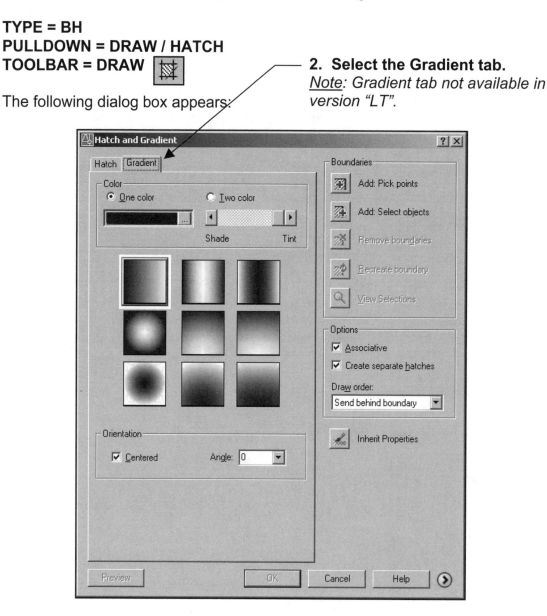

   The following dialog box appears:

   2. Select the Gradient tab.
   Note: Gradient tab not available in version "LT".

2. Select the Gradient tab.
3. Select the area for the gradient fill. (Pick Points or Select Objects)
4. Choose the Gradient settings. (Refer to page 15-7)
5. Preview
6. Accept or make changes.

GRADIENT FILLS continued...

ONE COLOR
Click the ... button to the right of the color swatch to open the Select Color dialog box.
Choose the color you want.

Use the "**Shade and Tint**" slider to choose the gradient range from lighter to darker.

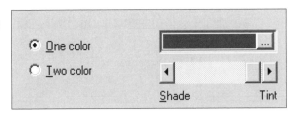

TWO COLOR
Click each ... button to choose a color.
When you choose two colors the transition is both from light to dark and from the first color to the second.

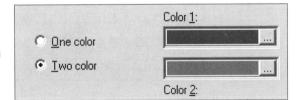

9 GRADIENT STYLES
Select one of the 9 gradient styles. (The selected style will have a white box around it.)

CENTER
Select "**Centered**", with a check mark in the box, to create a symmetrical fill.
Remove the check mark to move the "highlight" up and to the left.

ANGLE
Specify an angle for the "highlighted" area from the drop down list.
(The angle rotates counter clockwise, 0 is upper left corner.)

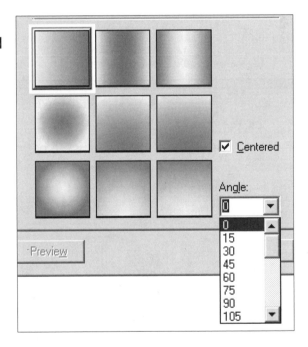

NOTES:
1. When you create a gradient, it appears in front of the boundary and sometimes obscures the object's outline.
To send the gradient hatch behind the boundary, specify the Draw Order.

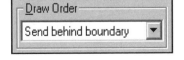

2. It is good drawing management to always place gradient fill on it's own layer.

3. You may also make gradient fill appear or disappear with the **FILL** command.
Type "**FILL**" **<enter>** on the command line. Then type "**ON**" or "**OFF**".
Select **VIEW / REGEN** to see the effect.

EDITING HATCH

HATCHEDIT allows you to edit an existing hatch pattern in the drawing.
You simply double click on the hatch pattern that you want to change and the hatch dialog box will appear. Make the changes, preview and accept.

1. Double click on the Hatch that you wish to edit, or select the Hatch Edit command using one of the following and then select the Hatch to edit:

TYPE = HE
PULLDOWN = MODIFY / OBJECT / HATCH
TOOLBAR = MODIFY II

2. Make the changes to the settings and preview. Your changes should be displayed.

3. Right click to accept or ESC to make additional changes.

TRIMMING HATCH

You may trim a hatch set just like any other object. But... the hatch set will no longer be associative. Meaning, if you change the shape of the boundary the hatch set will not change.

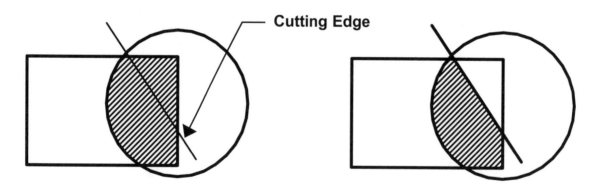

Cutting Edge

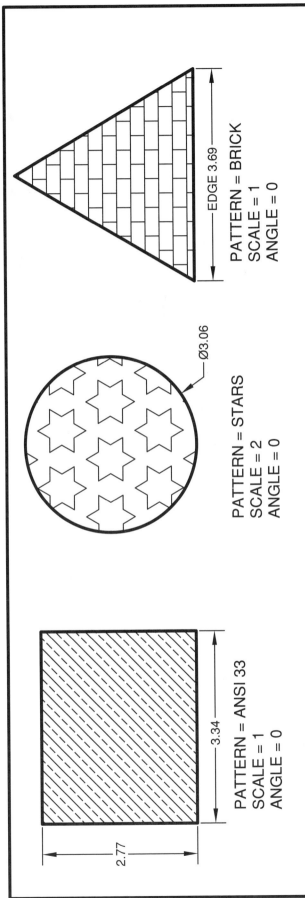

PATTERN = ANSI 33
SCALE = 1
ANGLE = 0

3.34

2.77

PATTERN = STARS
SCALE = 2
ANGLE = 0

Ø3.06

PATTERN = BRICK
SCALE = 1
ANGLE = 0

EDGE 3.69

INSTRUCTIONS:

1. Draw the **RECTANGLE, CIRCLE AND POLYGON** above approximately as shown. (Use layer = Object)
2. Select the **BHATCH** command. (Use Layer = Hatch)
3. Select Type: **PREDEFINED.**
4. Select the pattern listed under each object.
5. Select the **PICK POINTS** button.
   a. Place the cursor inside the hatch area and left click.
6. Right click and select Preview.
   a. Press ESC to make changes.
   b. Press <enter> to accept.
7. If not ok, make changes to the hatch options and preview again.
8. If ok, press <enter> to accept.

NAME	USING HATCH PREDEFINED PATTERNS	EX-15A

INSTRUCTIONS:

1. Follow the directions above.
2. Save as: **EX-15A and plot.**

EXERCISE 15A

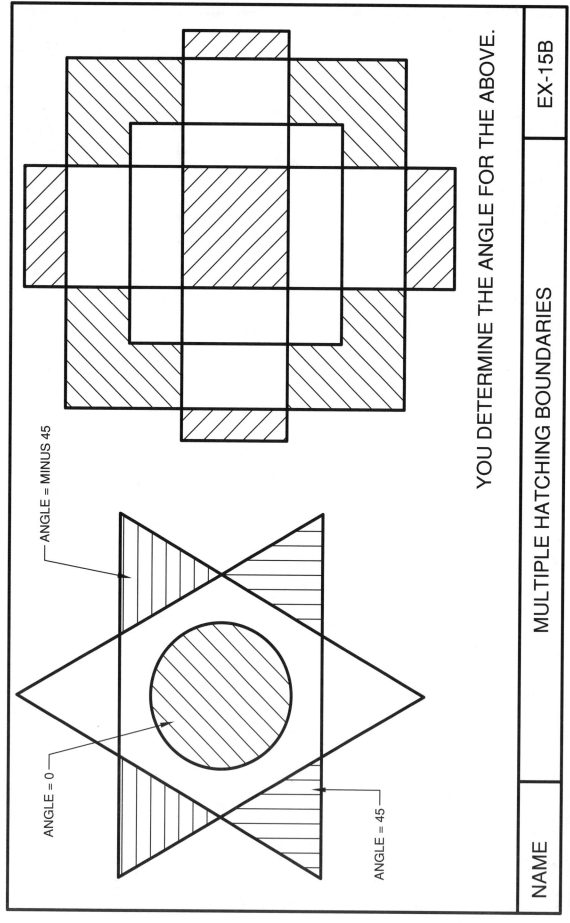

YOU DETERMINE THE ANGLE FOR THE ABOVE.

ANGLE = MINUS 45

ANGLE = 0

ANGLE = 45

MULTIPLE HATCHING BOUNDARIES

EX-15B

EXERCISE 15B

NAME

INSTRUCTIONS:

1. Draw the POLYGONS, CIRCLE AND RECTANGLES above.
2. HATCH as shown. (Use step by step procedure on EX-15A)
3. PATTERN = ANSI 31   SCALE = 2   ANGLE = noted above
4. Save as: EX-15B and plot.

EXERCISE 15C

EX-15C

HATCHING AROUND TEXT AUTOMATICALLY

TEXT

NAME

INSTRUCTIONS:

1. Draw the RECTANGLE above (Layer = Object)
2. Place 1.00 Ht. text in the middle of the Rectangle.   (Layer = Txt-Hvy)
3. Select options: Pattern = ANSI 31   Scale = 2   Angle = 0
4. Click on the SELECT OBJECTS button and select the Rectangle and Text.
5. Preview and <enter> if OK.
6. Save as: **EX-15C and plot.**

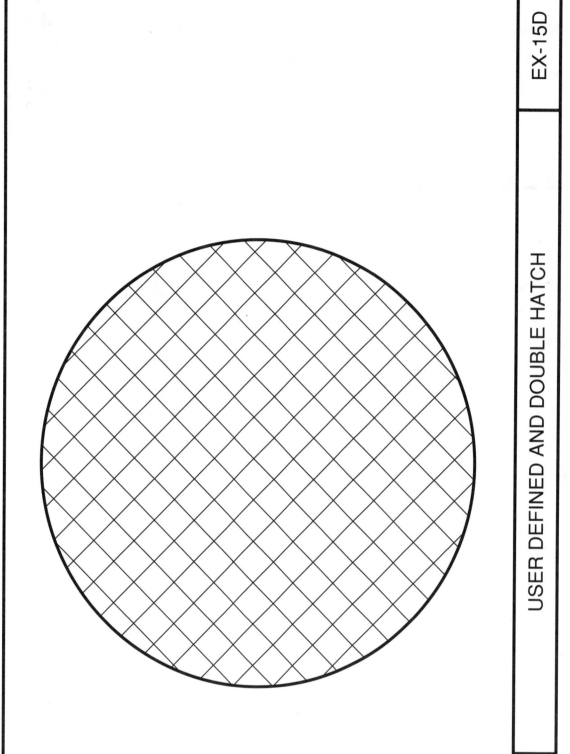

EX-15D

USER DEFINED AND DOUBLE HATCH

EXERCISE 15D

NAME

INSTRUCTIONS:

1. Draw the CIRCLE above (Layer = Object)
2. Draw the hatch using Type: **USER DEFINED** (Layer = Hatch)
3. Select options: Spacing = .50   Angle = 45
4. Select the **DOUBLE** box.
5. Select **PICK POINTS** button and click inside of the circle.
6. Preview, OK and Save as: **EX-15D and plot.**

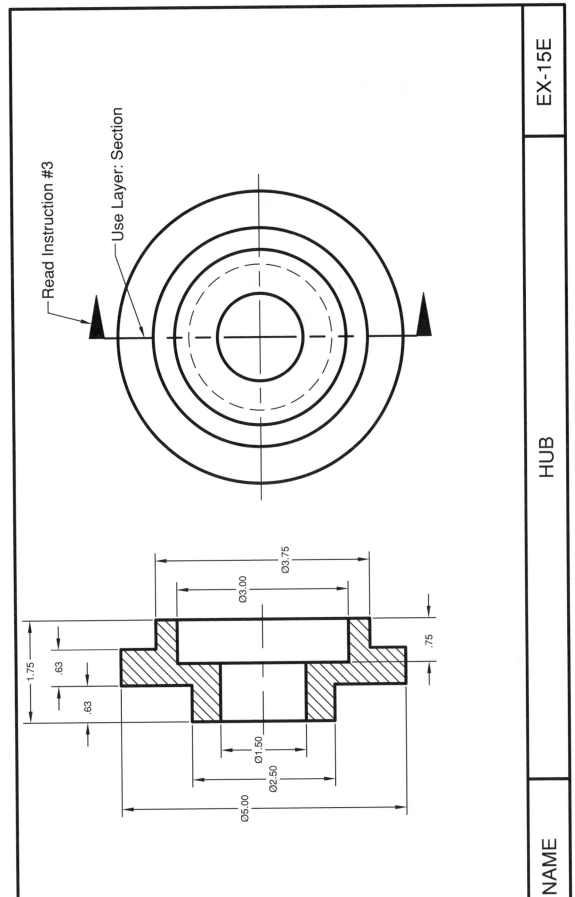

Read Instruction #3

Use Layer: Section

Ø3.75

Ø3.00

1.75

.63

.63

.75

Ø1.50

Ø2.50

Ø5.00

EXERCISE 15E

EX-15E

HUB

NAME

INSTRUCTIONS:

1. Draw the HUB above (Layer = Object)   Do not dimension
2. Draw the hatch using Pattern = ANSI 31   Scale = 1 Angle = 0 (Layer = Hatch)
3. Draw the section arrow, approximately as shown, then fill using "Solid" hatch. (Layer = Hatch)
4. Save as: EX-15E and plot

EX-15F

EDITING HATCH PATTERNS

NAME

EXERCISE 15F

INSTRUCTIONS:

1. Open drawing 15B and change the exisitng hatch pattern to SOLID.
2. Select MODIFY / OBJECTS / HATCH or double click in each area.
3. Select the hatch sets to be edited.
4. Select hatch pattern SOLID, PREVIEW, then <enter>, if OK.
5. Save as EX-15F and plot.

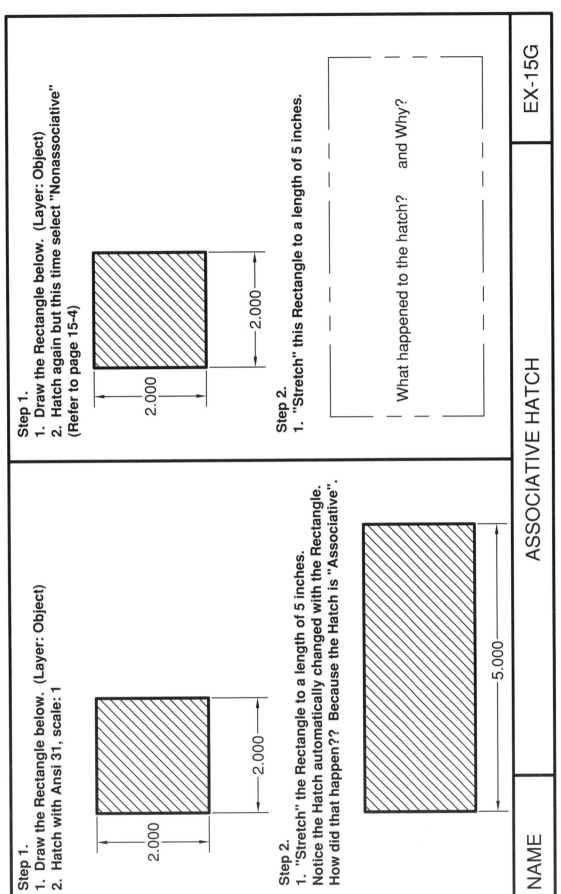

EXERCISE 15G

EX-15G

ASSOCIATIVE HATCH

Step 1.
1. Draw the Rectangle below. (Layer: Object)
2. Hatch again but this time select "Nonassociative"
(Refer to page 15-4)

2.000

2.000

Step 2.
1. "Stretch" this Rectangle to a length of 5 inches.

What happened to the hatch? and Why?

Step 1.
1. Draw the Rectangle below. (Layer: Object)
2. Hatch with Ansi 31, scale: 1

2.000

2.000

Step 2.
1. "Stretch" the Rectangle to a length of 5 inches.
Notice the Hatch automatically changed with the Rectangle.
How did that happen?? Because the Hatch is "Associative".

5.000

NAME

INSTRUCTIONS:
1. Follow the steps above.
2. Save as EX 15G <u>do not</u> Plot.

Step 1.

1. Draw the 2 Rectangles shown below. (Layer: Object)
   (Size is important   3 X 2)
2. Hatch the first Rectangle as follows:
   Pattern = Net   Scale = 2   Angle = 0

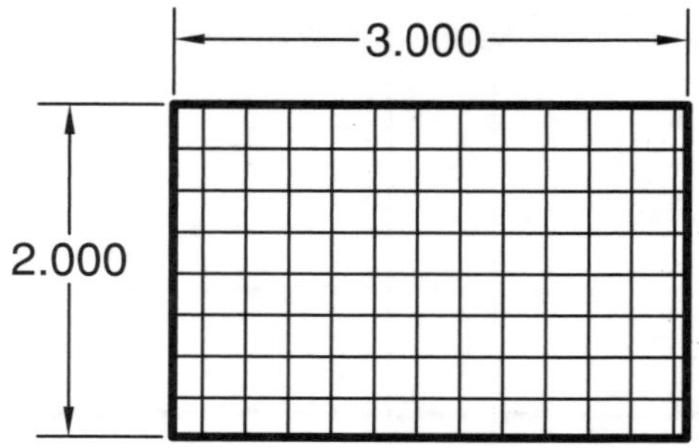

3.000

2.000

Step 2.

1. Hatch the second Rectangle using the same settings
   but this time select:
   a. Specified Origin
   b. Default to boundary extents
   c. Bottom left

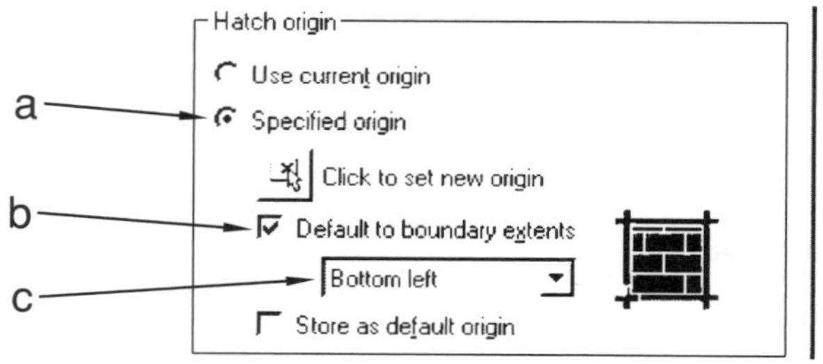

Notice that the hatch pattern aligned with the
"Bottom left" corner.
Compare this hatch with the one above.

Go back to EX 15A and experiment with Stars and Brick

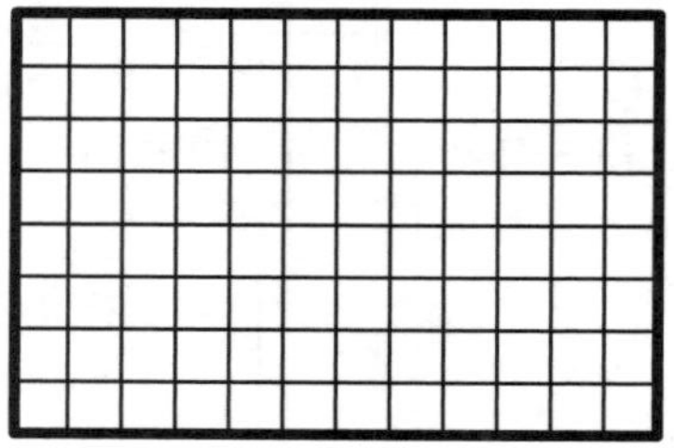

Hatch origin

⌒ Use current origin

⌖ Specified origin

[#] Click to set new origin

☑ Default to boundary extents

[Bottom left ▸]

☐ Store as default origin

a

b

c

NAME

SPECIFY THE HATCH ORIGIN

EX-15H

INSTRUCTIONS:

1. Follow the steps above.
2. Save as EX 15H do not Plot.

After completing this lesson, you will be able to:

1. Understand the importance of True Associative dimensioning.
2. Use Grips.
3. Add Linear, Baseline and Continued dimensions to your drawing.
4. Control the appearance of dimensions.
5. Create a New Dimension Style.
6. Compare two Dimension Styles.
7. Ignore Hatch Objects

LESSON 16

DIMENSIONING

Dimensioning is basically easy, but as always, there are many options to learn. As a result, I have divided the dimensioning process into 5 lessons. (Lessons 16 through 20) So relax and just take it one lesson at a time.

In this Lesson you will learn how to create a dimension style and how to create horizontal and vertical dimensions. But first you need to understand AutoCAD's true associative dimensioning setting.

True Associative

True Associative Dimensioning means that the dimensions are actually attached to the objects that they dimension. If you move the object, the dimension will move with it. If you scale or stretch the object, the dimension text value will change also. (Note: It is not parametric. This means, you can not change the dimension text value and expect the object to change.)

True Associative Dimensioning can be set to ON, OFF or Exploded.

I strongly suggest that you keep true associative dimensioning on. It is truly a very powerful feature and will make editing the objects and dimensions much easier.

On = Dimensions are truly associative. The dimensions are associated to the objects and will change if the object is changed. (Setting 2)

Off = Dimensions are non-associative. The dimensions are not associated to the objects and will not change if the object changes. (Setting 1)

Exploded = Dimensions are not associated to the objects and are totally separate objects (arrows, lines and text). (Setting 0)

How to turn Associative Dimensioning On or OFF.

There are 2 methods:

Method 1.
On the command line type: *dimassoc <enter>,*
then enter the number *2, 1* or *0 <enter>.*

Method 2.
Select **Tools / Options / User Preferences tab.**

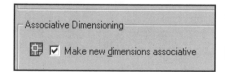

Checked box = On

Unchecked box = Off

Note: You can not set Exploded here.

The Default setting is "2" or "ON", but check to make sure.

When you first enter AutoCAD, the True Associative dimensioning feature is "ON". This setting is saved with each individual drawing. It is not a system setting for all drawings. This means, when you open a drawing it is important to check the **dimassoc** command to verify that True Associative dimensioning is ON. Especially if you open a drawing created with an AutoCAD release previous to 2002.

How to Re-associate a dimension.

If a dimension was created with a previous version of AutoCAD or True Associative dimensioning was turned off, you may use the **dimreassociate** command to change the non-associative dimensions into associative dimensions.
(You may use the **Tools / Inquiry / List** command to determine whether a dimension is associative or non-associative.)

1. Select **Dimension / Reassociate Dimensions**. (No icon available.)

   Command: _dimreassociate
   Select dimensions to reassociate ...

2. Select objects: *select the dimension to be reassociated.*

3. Select objects: *select more dimensions or <enter> to stop.*

4. Specify first extension line origin or [Select object] <next>: *an "X" will mark the first extension line; use object snap to select the exact location of the extension line point. (If the "X" has a box around it, just press <enter>.)*

5. Specify second extension line origin <next>: *the "X" will move to another location. Use object snap or <enter>.*

6. *Continue until all extension line points are selected.*

Note: You must use object snap to specify the exact location for the extension lines.

Regenerating Associative dimensions

Sometimes after panning and zooming, the associative dimensions seem to be floating or not following the object. The **DIMREGEN** command will move the associative dimensions back into their correct location.

*You must type **dimregen <enter>** on the command line. Sorry, AutoCAD did not provide an icon or a pull-down menu.*

GRIPS

Grips are little boxes that appear if you select an object when no command is in use. Grips must be enabled by typing "grips" <enter> then 1 <enter> on the command line or selecting the "enable grips" box in the **TOOLS / OPTIONS / SELECTION** dialog box.

Grips can be used to quickly edit objects. You can move, copy, stretch, mirror, rotate, and scale objects using grips.

The following is a brief overview on how to use three of the most frequently used options. Grips have many more options and if you like the example below, you should research them further in the AutoCAD help menu.

1. Select the object (no command can be in use while using grips)
2. Select one of the **blue** grips. It will turn to "**red**". This indicates that it is "**hot**". The "**Hot**" grip is the **basepoint**.
3. The editing modes will be displayed on the command line. You may cycle through these modes by pressing the SPACEBAR or ENTER key or use the shortcut menu.
4. <u>After editing you must press the ESC key to deactivate the grips on that object.</u>

Selecting a grip:
When you select a grip it becomes "HOT".

Hot grip

Moving an object:
1. Select the object.
2. Select the grip in the middle of the object.
3. Move the cursor to the new location.
4. Left click.

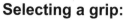

Move

Coping an object:
1. Select the object.
2. Select the grip.
3. Select the COPY option from the Stretch mode.
4. Move the cursor to the new location for the copy(s) and left click.
Note: Grips will allow you to continue making copies until you press the ESC key to stop.

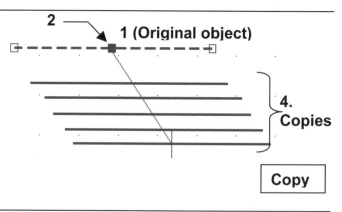

Copy

Stretch an object:
1. Select the object
2. Select the grip.
3. Move the cursor to stretch the object or type @X,Y

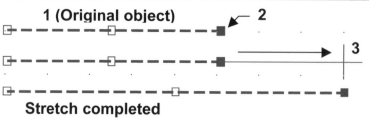

Stretch completed

Stretch

REMEMBER, press ESC key to deactivate the grips on an object.

LINEAR DIMENSIONING

First open the dimension toolbar. Using the toolbar icons to select the dimension commands is the most efficient method. (Refer to page 2-9 to open toolbar.)

<u>Linear dimensioning</u> allows you to create horizontal and vertical dimensions.

1. Select the **LINEAR** command using the icon or Dimension / Linear.
   Command:__dimlinear
2. Specify first extension line origin or <select object>: ***snap to first extension line origin (P1).***
3. Specify second extension line origin: ***snap to second extension line origin (P2).***
4. Specify dimension line location or [Mtext/Text/Angle/Horizontal/Vertical/Rotated]: ***select where you want the dimension line placed (P3).***
   Dimension text = 2.000 *(The dimension text value will be displayed on the last line.)*

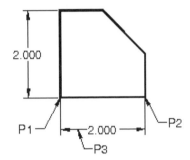

BASELINE DIMENSIONING

<u>Baseline dimensioning</u> allows you to establish a **baseline** for successive dimensions. The spacing between dimensions is automatic and should be set in dimension styles. A Baseline dimension must be used with an existing dimension. If you use Baseline dimensioning immediately after a Linear dimension, you do not have to specify the baseline origin.

1. Create a <u>linear</u> dimension first (1.400   P1 and P2).
2. Select the **BASELINE** command using the icon or Dimension / Baseline.
   Command: _dimbaseline
3. Specify a second extension line origin or [Undo/Select] <Select>: ***snap to the second extension line origin (P3).***
   Dimension text = 2.588
4. Specify a second extension line origin or [Undo/Select] <Select>: ***snap to P4.***
   Dimension text = 3.633
5. Specify a second extension line origin or [Undo/Select} <Select>: ***select <enter> twice to stop.***

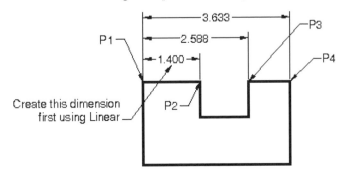

CONTINUE DIMENSIONING

Continue creates a series of dimensions in-line from an existing dimension. If you use the continue dimensioning immediately after a Linear dimension, you do not have to specify the continue extension origin.

1. Create a linear dimension first (1.400   P1 and P2).
2. Select the Continue command using the icon or Dimension / Continue.
   Command: _dimcontinue
3. Specify a second extension line origin or [Undo/Select] <Select>: ***snap to the second extension line origin (P3).***
   Dimension text = 1.421
4. Specify a second extension line origin or [Undo/Select] <Select>: ***snap to the second extension line origin (P4).***
   Dimension text = 1.364
5. Specify a second extension line origin or [Undo/Select] <Select>: ***press <enter> twice to stop.***

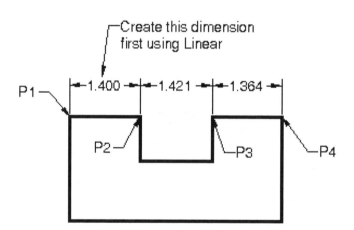

DIMENSION STYLES

Using the "Dimension Style Manager" you can change the appearance of the dimension features, such as length of arrowheads, size of the dimension text, etc. There are over 70 different settings.
You can also Create New, Modify, Override and Compare Dimension Styles. All of these are simple, by using the Dimension Style Manager described below.

Select the "Dimension Style Manager" using one of the following:

TYPE = DDIM
PULLDOWN = DIMENSION / STYLE
TOOLBAR = DIMENSION

The following dialog box will appear.

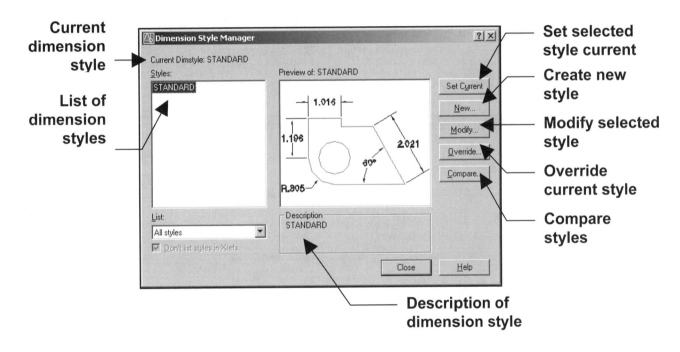

Current dimension style

List of dimension styles

Set selected style current

Create new style

Modify selected style

Override current style

Compare styles

Description of dimension style

Set Current   Select a style from the list of styles and select the **set current** button. (Only Standard is shown unless you have previously created other styles.)

New   Select this button to create a new style. When you select this button, the **Create New Dimension Style** dialog box is displayed. (See page 16-8.)

Modify   Selecting this button opens the **Modify dimension Style** dialog box which allows you to make changes to the current style. (See page 17-4.)

Override   An override is a temporary change to the current style. Selecting this button opens the Override Current Style dialog box. (See page17-5.)

Compare   Compares two styles. (See page 16-12.)

CREATING A NEW DIMENSION STYLE

When creating a new style you must start with an existing style, such as Standard. Next, assign it a new name, make the desired changes and when you select the OK button the new style will have been successfully created.

LET'S CREATE A NEW DIMENSION STYLE.

A dimension style is a group of settings that has been saved with a name you assign.

1. Open your drawing **BSIZE** and set **DIMASSOC** to **2.**
2. Select the **DIMENSION STYLE** command (Ref. page 16-7).
3. Select the **NEW** button (Ref. page 16-7).
4. Enter **CLASS STYLE** in the "New Style Name" box.
5. **"Start With:" box**: Start with the settings in the STANDARD style and then you will make some changes.
6. **"Use For:" box** is for creating "family" dimension styles and will be discussed later. For now, leave it set to "All dimensions".
7. Select the **CONTINUE** button.

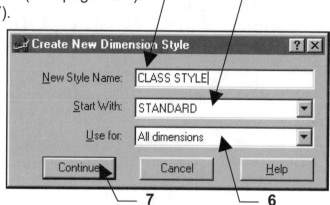

The "New Dimension Style: Class Style" dialog box will appear.

8. Select the **"Primary Units"** tab and make the changes shown below.

NOTE: REFER TO "APPENDIX B" FOR DESCRIPTIONS OF ALL DIMENSION SETTINGS

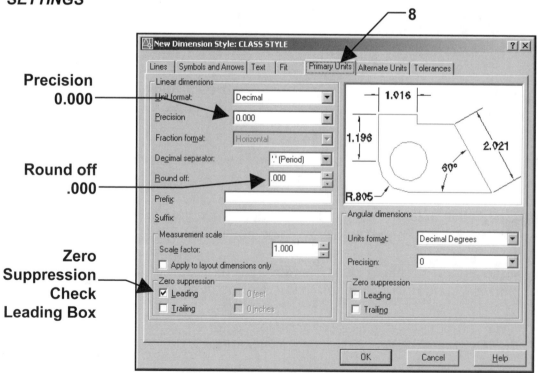

Precision
0.000

Round off
.000

Zero Suppression Check Leading Box

DO NOT SELECT THE OK BUTTON YET.

9. Select the **"Lines"** tab and make the changes shown below.

Baseline spacing .500

Extend beyond .125

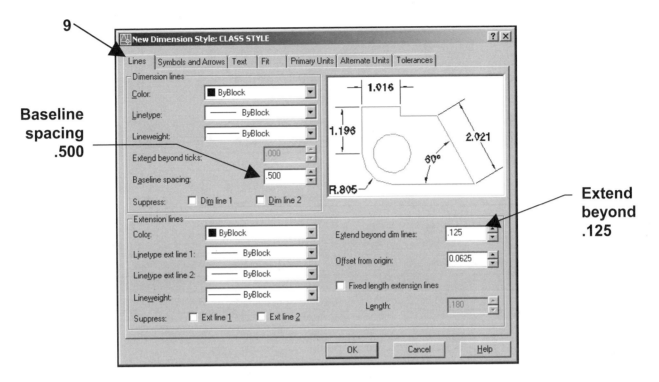

DO NOT SELECT THE OK BUTTON YET.

10. Select the **"Symbols and Arrows"** tab and make the changes shown below.

Arrow size .125

Center Marks Line

Size .120

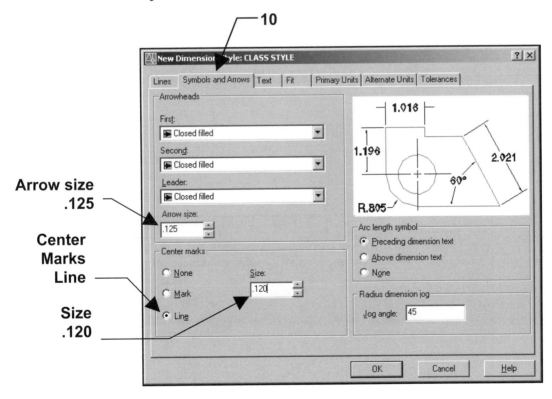

DO NOT SELECT THE OK BUTTON YET.

11. Select the **"Text"** tab.

11

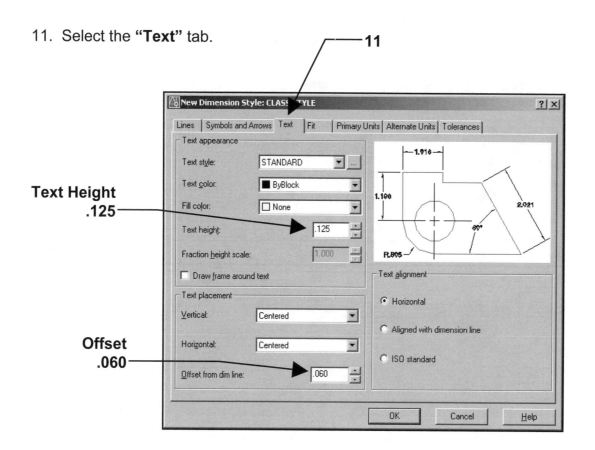

**Text Height
.125**

**Offset
.060**

DO NOT SELECT THE OK BUTTON YET.

12. Select the **"Fit"** tab.

12

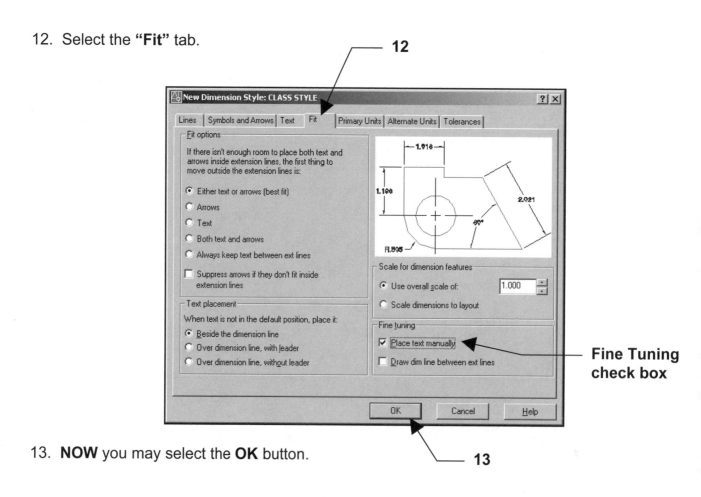

**Fine Tuning
check box**

13. **NOW** you may select the **OK** button.

13

14. *Your new style **"Class Style"** should be listed.* Select the **"Set Current"** button to make your new style "Class Style" the style that will be used.

Your new style is listed here.

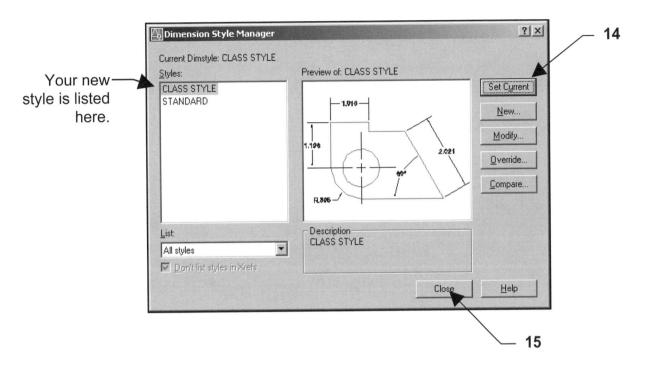

14

15

15. Select the **Close** button to **exit**.

16. Save your drawing as **BSIZE**.

Note: You have successfully created a new "Dimension Style" called "Class Style". This style will be saved in your BSIZE drawing after you save the drawing. It is important that you understand that this dimension style resides only in the BSIZE drawing. If you open another drawing, this dimension style will not be there.

COMPARE TWO DIMENSION STYLES

Sometimes it is useful to compare the settings of two styles. Compare will compare the two styles and list the differences.

LET'S COMPARE "CLASS STYLE" AND "STANDARD"

1. Select the Dimension Style command.
2. Select the **Compare** button.

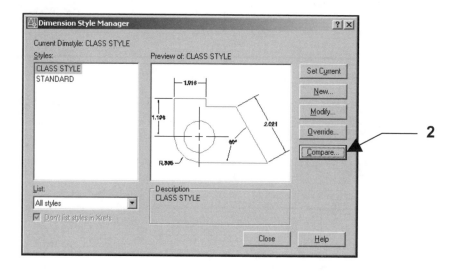

The "Compare Dimension Style" dialog box will appear.

3. Select **Class Style** in the **Compare** box.
4. Select **Standard** in the **With** box.
5. AutoCAD found differences and listed them.
6. Select "Close" button to exit.

Yours may be slightly different.

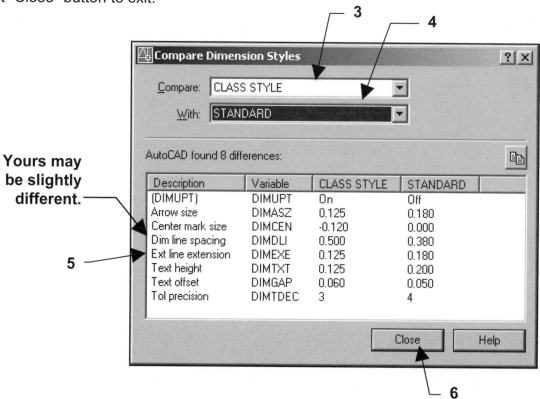

IGNORING HATCH OBJECTS

Occasionally, when you are dimensioning an object that has "Hatch Lines", your cursor will snap to the Hatch Line instead of the object that you want to dimension. To prevent this from occurring, select the option **"IGNORE HATCH OBJECTS"**.

EXAMPLE:

**Ignore Hatch Objects
<u>Not</u> Selected**

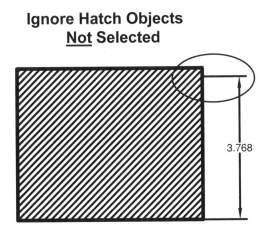

3.768

**Ignore Hatch Objects
Selected**

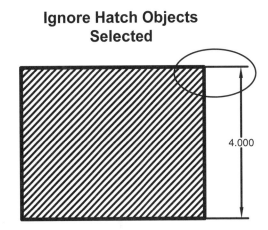

4.000

<u>How to select the option "**IGNORE HATCH OBJECTS**".</u>

1. Select **TOOLS / OPTIONS / DRAFTING**
2. Select "Ignore Hatch Objects" box.

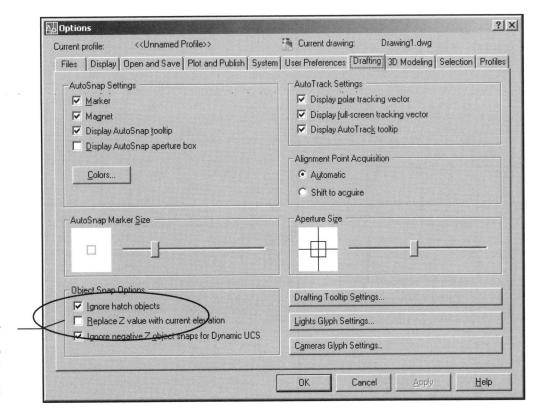

**Version LT
does not have
"Replace Z"
option**

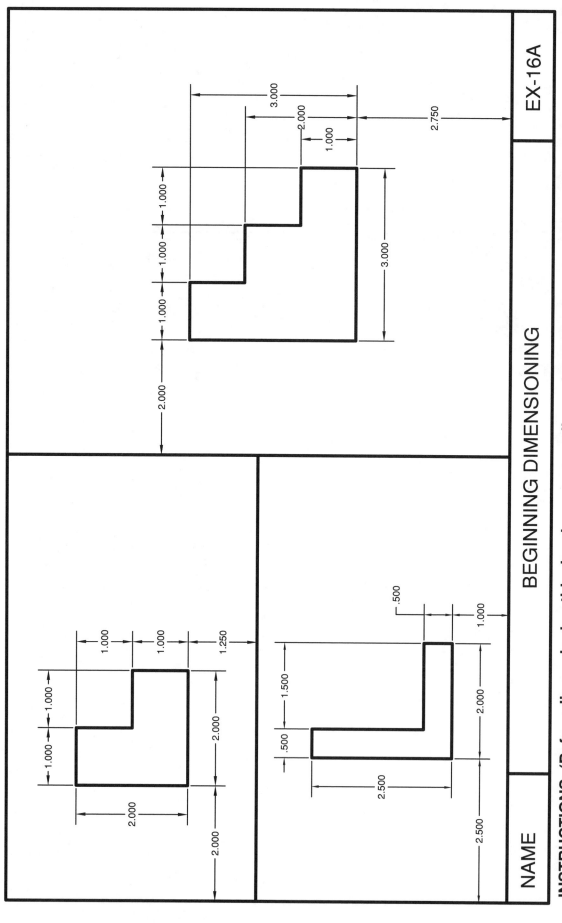

BEGINNING DIMENSIONING

EXERCISE 16A

EX-16A

NAME

INSTRUCTIONS: (Before dimensioning this drawing, create a dimension style in your Bsize master drawing. Follow the instructions on page 16-8 thru 11)

1. Draw the objects above. (Use Layer: Object)
2. Dimension as shown using Dimension Style: Class Style and Linear, Baseline and Continue. (Use Layer: Dim)
3. Try to duplicate the dimensions as shown.
4. Save as: **EX-16A**

16-14

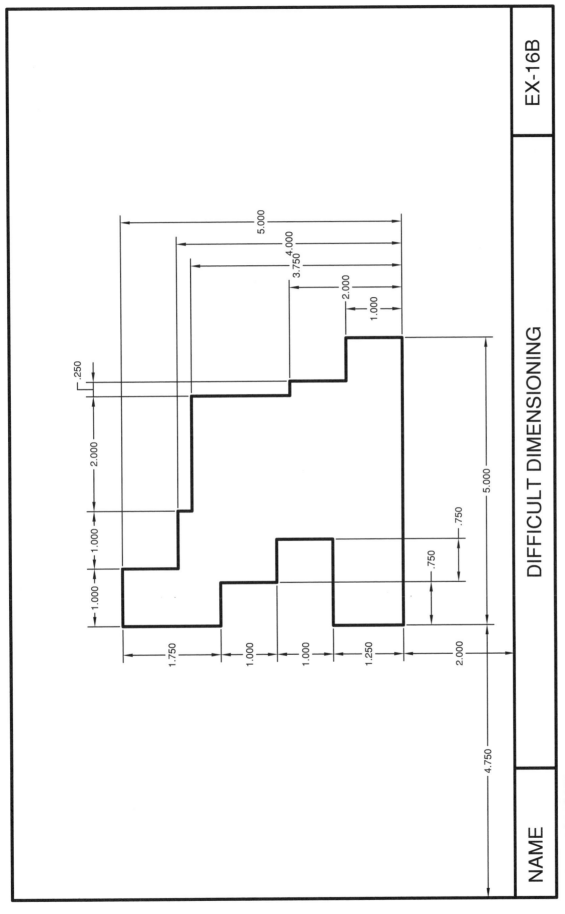

EXERCISE 16B

DIFFICULT DIMENSIONING

EX-16B

NAME

INSTRUCTIONS:

1. Draw the objects above. (Use Layer: Object)
2. Note: The OFFSET command would be more efficient than coordinate input.
3. Dimension as shown using Dimension Style: Class Style and Linear,
   Baseline and Continue. (Use Layer: Dim)
4. Try to duplicate the dimensions as shown.
5. Save as: **EX-16B**

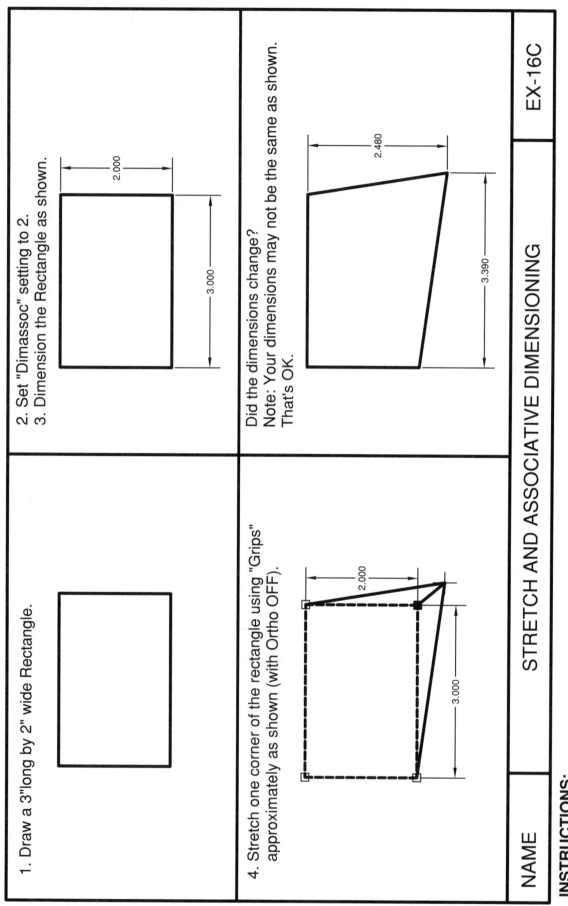

EXERCISE 16C

1. Draw a 3"long by 2" wide Rectangle.

2. Set "Dimassoc" setting to 2.
3. Dimension the Rectangle as shown.

2.000

3.000

4. Stretch one corner of the rectangle using "Grips" approximately as shown (with Ortho OFF).

2.000

3.000

Did the dimensions change?
Note: Your dimensions may not be the same as shown. That's OK.

2.480

3.390

NAME

STRETCH AND ASSOCIATIVE DIMENSIONING

EX-16C

INSTRUCTIONS:

1. Follow the instructions above.
2. Do not divide your drawing into 4 sections. Just draw one rectangle in the middle of your drawing area and follow the instructions.
3. Save as: **EX-16C.**

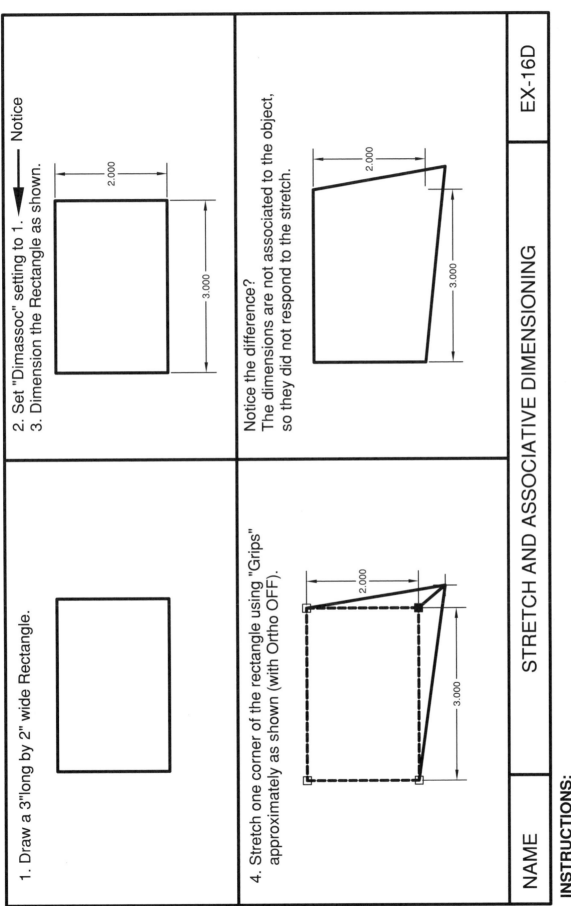

1. Draw a 3"long by 2" wide Rectangle.

2. Set "Dimassoc" setting to 1. ⟵ Notice
3. Dimension the Rectangle as shown.

2.000

3.000

4. Stretch one corner of the rectangle using "Grips" approximately as shown (with Ortho OFF).

2.000

3.000

Notice the difference?
The dimensions are not associated to the object, so they did not respond to the stretch.

2.000

3.000

NAME	STRETCH AND ASSOCIATIVE DIMENSIONING	EX-16D

EXERCISE 16D

INSTRUCTIONS:

1. Follow the instructions above.
2. Do not divide your drawing into 4 sections.
   Just draw one rectangle in the middle of your
   drawing area and follow the instructions.
3. Save as: **EX-16D.**

NOTES:

LEARNING OBJECTIVES

After completing this lesson, you will be able to:

1. Edit Dimension text values
2. Edit the Dimension position.
3. Modify an entire Dimension Style.
4. Override a Dimension Style.
5. Edit a Dimension using Properties Palette.

LESSON 17

EDITING DIMENSION TEXT VALUES

Sometimes you need to modify the dimension value text. You may add a symbol, a note or even change the text of an existing dimension. The following describes 2 methods.

<u>Example</u>: Add the word "Max." to the existing dimension value text.

Before Editing   **After Editing**

   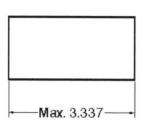

Method 1 (Properties Palette).
1. Double click on the dimension that you want to change.
   The Properties Palette will appear.
2. Scroll down to **Text / Text override**
   (Notice that the actual <u>measurement</u> is directly above it.)
3. Type the new text (Max) and **< >** and press <enter>
   (< > represents the associative text value)

Rotated Dimension	
Fill color	None
Fractional t...	Horizontal
Text color	■ ByBlock
Text height	.125
Text offset	.060
Text outsid...	Off
Text pos hor	Centered
Text pos vert	Centered
Text style	STANDARD
Text inside ...	Off
Text positio...	8.191
Text positio...	3.196
Text rotation	0
Measurement	3.337
Text override	Max <>

Fit

Primary Units

Alternate Units

Tolerances

Specifies the text string of the dimension (overrides Measurement string)

> **Important:**
> **If you do not use < > the dimension will no longer be Associative.**

Method 2 (ddedit).
1. Type on the Command line: **ed** <enter>   (This is the **ddedit** command)
2. Select the dimension you want to edit.
3. The "In-Place Text Editor" will appear.

Associative Dimension
If the dimension is <u>Associative</u> the dimension text will appear highlighted.

You may add text in front or behind the dimension text and it will remain Associative. Be careful not to disturb the dimension value text.

— Max 9.815 —

Non Associative Dimension
If the dimension value has been changed or exploded it will appear with a gray background and is not Associative.

— 9.815 Max —

4. Make the change.
5. Select the OK button.

EDITING THE DIMENSION POSITION

Sometimes dimensions are too close and you would like to stagger the text or you need to move an entire dimension to a new location, such as the examples below.

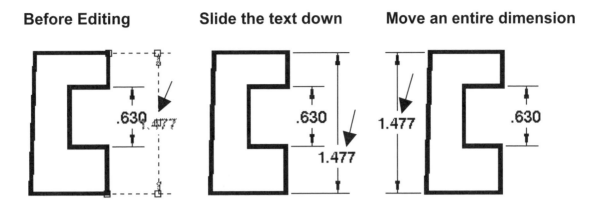

| Before Editing | Slide the text down | Move an entire dimension |

Editing the position is easy.
1. Select the dimension that you want to change. (Grips will appear)
2. Select the middle grip. (It will become red and active, hot)
3. Move the cursor and the dimension will respond.
4. Press the left mouse button to place the dimension in the new location.

ADDITIONAL EDITING OPTIONS USING THE SHORTCUT MENU

1. Select the dimension that you want to change.
2. Press the Right Mouse button.
3. Select "Dimension Text Position" from the shortcut menu shown to the right.
4. A sub-menu appears with many editing options.

Experiment with these options. They will be very helpful in the future.

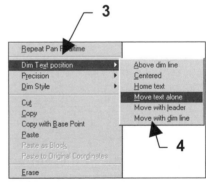

**Note: This command will not work with <u>Exploded</u> dimensions.
You will have to use the "<u>Stretch</u>" command to achieve the same results.**

MODIFY AN <u>ENTIRE</u> DIMENSION STYLE

After you have created a Dimension Style, you may find that you have changed your mind about some of the settings. You can easily change the entire Style by using the "Modify" button in the Dimension style Manager dialog box. This will not only change the Style for future use, but it will also <u>update dimensions already in the drawing</u>.

Note: if you do not want to update the dimensions already in the drawing, but want to make a change to the next dimension drawn, refer to Override, page 17-5

1. Select the Dimension Style command. (Refer to page 16-7)

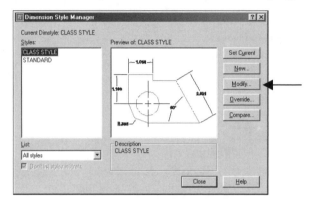

2. Select the Modify button from the Dimension Style Manager dialog box.

3. Make the desired changes to the settings.

4. Select the OK button.

5. Select the Close button.

Now look at your drawing. Have your dimensions updated?

Note: If some of the dimensions have not changed:
1. Select Dimension / Update
2. Select the dimension(s) you wish to update and press <enter>

Note: This feature <u>will not</u> change dimensions that have been modified or exploded.

OVERRIDE A DIMENSION STYLE

A dimension Override is a **temporary** change to the dimension settings.
An override **will not affect existing dimensions**. It will affect **new dimensions only**.

*Use this option when you want your **next dimension** just a little bit different but you don't want to create a whole new dimension style and you don't want the existing dimensions to change either.*

1. Select **Format / Dimension Style.**

2. Select the "**Style**" you want to override. (such as Class Style)

3. Select the **Override** button.

4. Make the desired changes to the settings.

5. Select the OK button.

6. Confirm the Override
   a. Look at the List of styles. Under the Style name, a sub heading of "Style overrides" should be displayed.
   b. The description box should display the style name and the override settings.

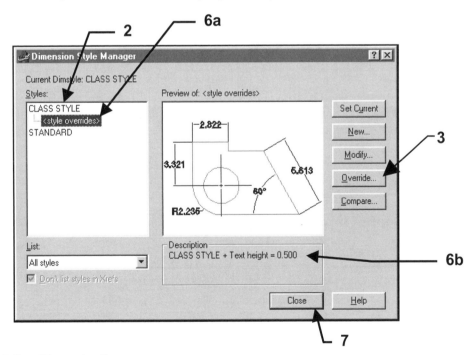

7. Select the Close button.

If you want to return to style "**Class Style**", select "Class Style" and then select the "**Set Current**" button. Each time you select a different style, you must select the **Set Current** button to activate it.

EDITING A DIMENSION using PROPERTIES PALETTE

Sometimes you would like to modify the settings of an **individual existing** dimension. This can be achieved using the Modify Properties command.

1. Double click on the dimension that you wish to change.

   The Properties Palette will appear.

2. Select and change the desired settings.

   Example: Change the dimension text height or size of the Arrow.

3. Press <enter> . *(The change should have taken effect)*

4. Press the <esc> key

Note: The dimension will remain Associative.

Note: This process <u>will not</u> change <u>Exploded</u> dimensions.

EXERCISE 17A

NAME	OVERRIDE	EX-17A

INSTRUCTIONS:

1. Draw the 6" Long by 4" Wide Rectangle above. (Use Layer: Object)
2. Use dimension style "Class Style" for dimension A. (Use Layer: Dim)
3. Use OVERRIDE for dimension B. Change the setting for Text height to .500.
4. Save as: **EX-17A and plot.**

EXERCISE 17B

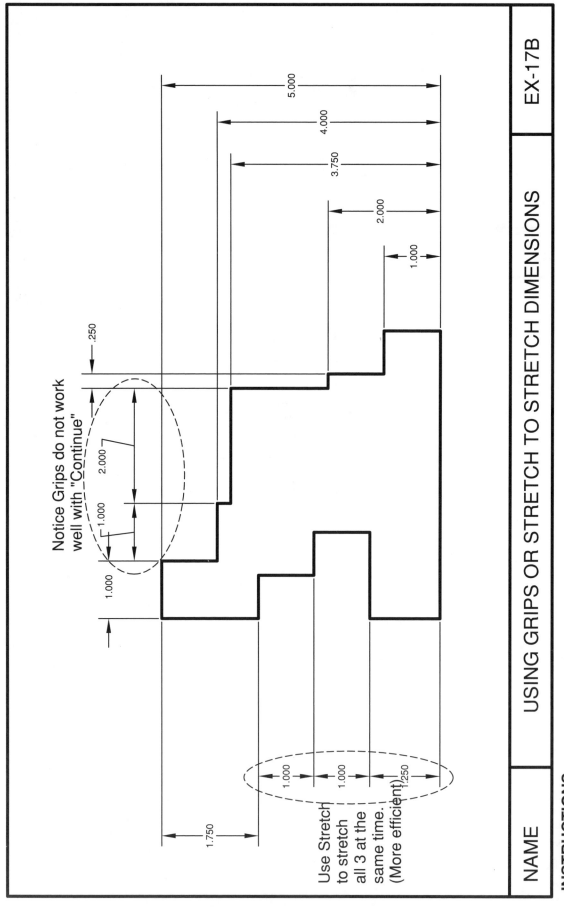

Notice Grips do not work well with "Continue"

Use Stretch to stretch all 3 at the same time. (More efficient)

NAME	USING GRIPS OR STRETCH TO STRETCH DIMENSIONS	EX-17B

INSTRUCTIONS:

1. Open drawing 16B.

2. Try using Grips to stretch the dimensions as shown.
You may find that the STRETCH command works better in some cases.

3. Save as: **EX-17B and plot.**

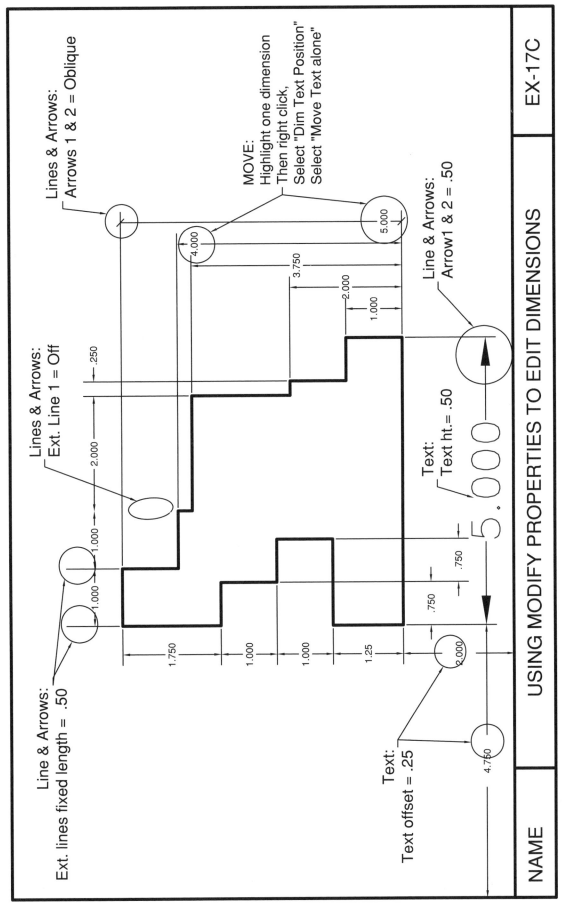

EXERCISE 17C

Lines & Arrows:
Arrows 1 & 2 = Oblique

MOVE:
Highlight one dimension
Then right click,
Select "Dim Text Position"
Select "Move Text alone"

Line & Arrows:
Arrows 1 & 2 = .50

Lines & Arrows:
Ext. Line 1 = Off

Line & Arrows:
Ext. lines fixed length = .50

Text:
Text ht.= .50

Text:
Text offset = .25

| NAME | USING MODIFY PROPERTIES TO EDIT DIMENSIONS | EX-17C |

INSTRUCTIONS:

1. Open Drawing 16B.
2. Edit the dimensions using: MODIFY / PROPERTIES.
3. Save as: **EX-17C and plot.**

NOTES:

LEARNING OBJECTIVES

After completing this lesson, you will be able to:

1. Add Diameter, Radial and Angular dimensions to your drawing.
2. Draw Center Marks.
3. Control the size and appearance of the Center Marks.
4. Understand the need for Sub-Styles.
5. Create a Sub-Style.

LESSON 18

RADIAL DIMENSIONING

DIAMETER DIMENSIONING

The **DIAMETER** dimensioning command should be used when dimensioning circles and arcs of <u>more than 180 degrees</u>.

<u>Center marks</u> are automatically drawn as you use the diameter dimensioning command. If the circle already has a center mark or you do not want a center mark, set the center mark setting to **NONE** (Dimension Style / Symbols and Arrows tab) before using Diameter dimensioning.

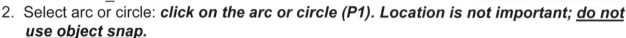

1. Select the **Dimension / Diameter** command or
   Command: _dimdiameter
2. Select arc or circle: *click on the arc or circle (P1). Location is not important; <u>do not use object snap.</u>*
   Dimension text = *the diameter will be displayed here.*
3. Specify dimension line location or [Mtext/Text/Angle]: *place dimension text location (P2).*

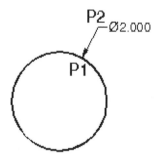

If you would like to **FLIP** the direction of the arrowhead:
1. Select the dimension
2. Press the right mouse button.
3. Select "**Flip Arrow**" from the menu.

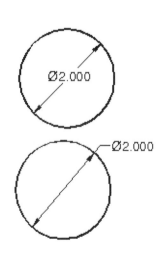

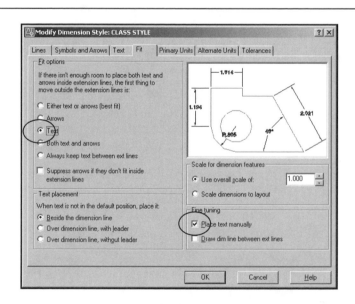

If you would like your Diameter dimensions to appear as shown in the two examples above, you must change the "**Fit Options**" and "**Fine Tuning**" in the **Fit tab** in your Dimension Style.

RADIAL DIMENSIONING (continued)

RADIUS DIMENSIONING

The **RADIUS** dimensioning command should be used when dimensioning arcs of <u>LESS than 180 degrees.</u>

<u>Center Marks</u> are automatically drawn as you use the **RADIUS** dimensioning command. If the circle already has a center mark, set the center mark to **NONE** (in the Dimension Style) before using RADIUS dimensioning. Or create a "sub-style" as shown on page 18-6.

1. Select the **Dimension / Radius** command or
   Command: _dimradius
2. Select arc or circle: *click on the arc (P1). Location is not important, do not use snap.*
   Dimension text = *the radius will be displayed here*
3. Specify dimension line location or [Mtext/Text/Angle]: *place dim text location (P2)*

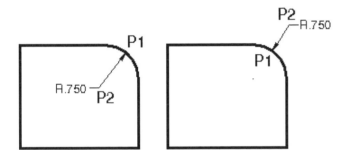

If you would like to **FLIP** the direction of the arrowhead:
1. Select the dimension
2. Press the right mouse button.
3. Select "**Flip Arrow**" from the menu.

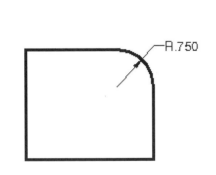

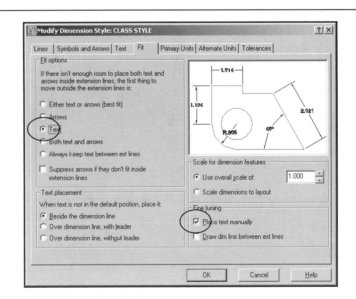

If you would like your Radius dimension to appear as shown in the example immediately above, you must change the "**Fit Options**" and "**Fine Tuning**" in the **Fit tab** in your Dimension Style.

ANGULAR DIMENSIONING

The **ANGULAR** dimension command is used to create an angular dimension between two lines that form an angle. All that is necessary is the selection of the two lines and the location for the dimension text.

The **degree symbol** is automatically added as the dimension is created.

1. Select the **ANGULAR** command.
   Command: _dimangular
2. Select arc, circle, line, or <specify vertex>: **click on the first line that forms the angle (P1) location is not important, do not use snap.**
3. Select second line: **click on the second line that forms the angle (P2)**
4. Specify dimension arc line location or [Mtext/Text/Angle]: **place dimension text location**
   Dimension text = **angle will be displayed here**

Any of the 4 angular dimensions shown below can be created by clicking on the 2 lines (P1 and P2) that form the angle.

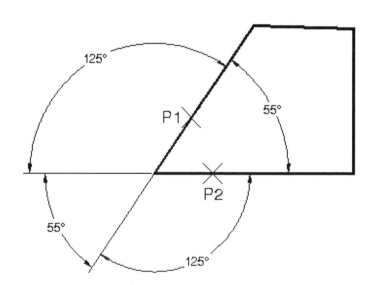

CENTERMARK

CENTERMARKS can ONLY be drawn with circular objects like Circles and Arcs.
You set the size and type.

The Center Mark has three types, **None, Mark** and **Line** as shown below.

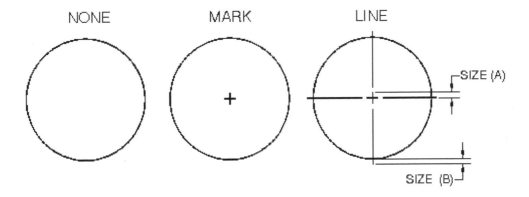

What does "SIZE" mean?

The **size** setting determines both, (A) the length of half of the intersection line and
(B) the length extending beyond the circle. (See above)

Where do you set the CENTERMARK "TYPE" and "SIZE"

1. Select the **Dimension Style** command.
2. Select: **New, Modify, or Override.**
3. Select: **SYMBOLS and ARROWS** tab
4. Select the **Center mark type**
5. Set the **Size.**

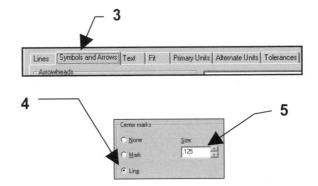

To draw a CENTER MARK

1. Select the **Center mark** command using one of the following:

> **TYPING = DIMCENTER or DCE**
> **PULLDOWN = DIMENSION / CENTERMARK**
> **TOOLBAR = DIMENSION**

2. Select arc or circle: *select the arc or circle with the cursor.*

CREATING A DIMENSION SUB-STYLE

Sometimes when using a dimension style, you would like the Linear, Angular, Diameter and Radius dimensions to have different settings. But you want them to use the same dimension style. To achieve this, you must create a "sub-style".

Sub-styles have also been called "children" of the "Parent" dimension style. As a result, they form a family.

A Sub-style is permanent, unlike the Override command, which is temporary.

LET'S CREATE A SUB-STYLE FOR RADIUS.

We will set the center mark to None for the Radius command only.
The Diameter command center mark will not change.

1. Open your **BSIZE** drawing.

2. Select the **DIMENSION / STYLE** command.

3. Select **"Class Style"** from the Style List.

4. Select the **NEW** button.

5. Change the "Use for" to:
   Radius Dimensions

6. Select **Continue**

7. Select the **Symbols and Arrows** tab.

8. Change the "Center Marks for Circles" Type: **NONE**

9. Select the **OK** button

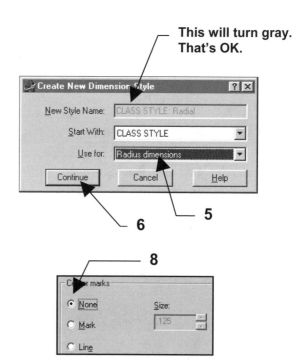

This will turn gray.
That's OK.

You have now created a sub-style that will automatically override the basic "Class Style" whenever you use the Radius command.

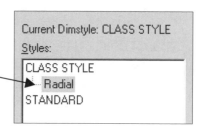

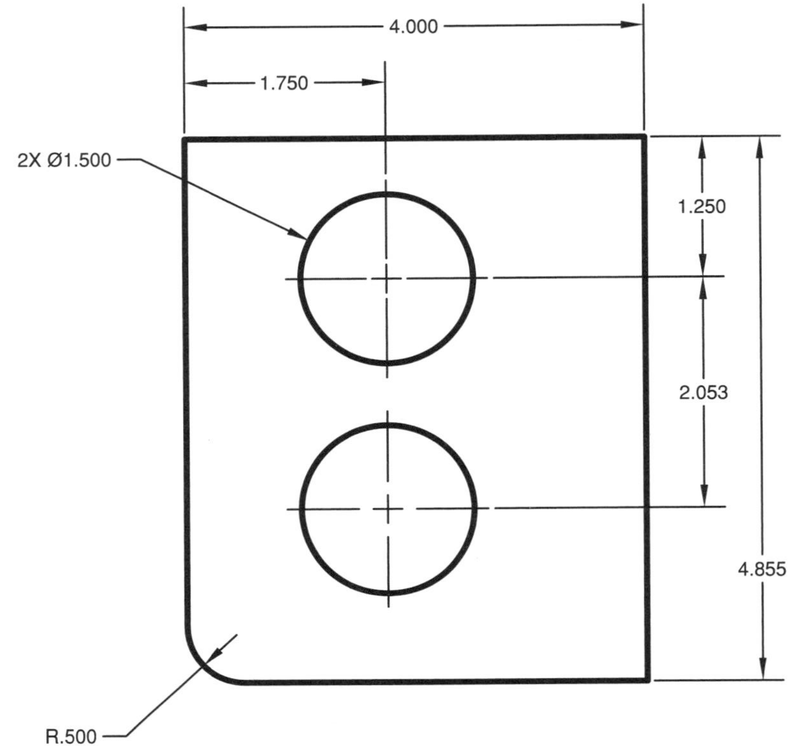

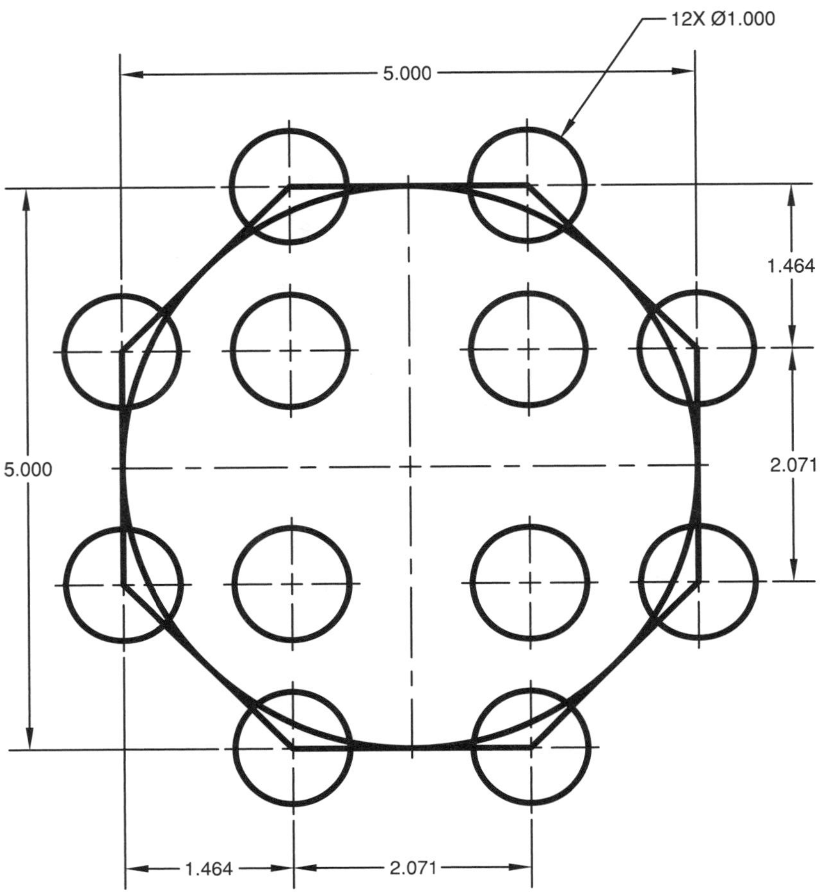

Top drawing dimensions
12X Ø1.000
1.464
2.071
5.000
5.000
2.071
1.464

Bottom drawing dimensions
2X Ø1.500
1.250
2.053
4.855
4.000
1.750
R.500

EX-18A

DIAMETER AND LINEAR DIMENSIONING

EXERCISE 18A

NAME

INSTRUCTIONS:

1. Draw the objects above.
2. Dimension as shown using LINEAR and DIAMETER dimensioning.
3. NOTE: Dimension the Diameter first, then add the center marks to the remaining circles. Draw the Linear dimensions last so you can snap to the endpoint of the center marks.
4. Save as: **EX-18A and plot.**

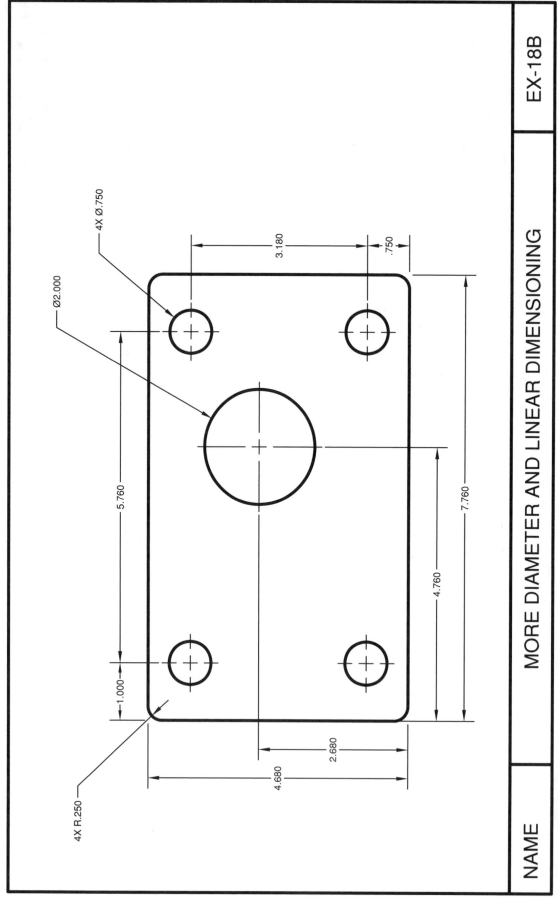

EX-18B

MORE DIAMETER AND LINEAR DIMENSIONING

EXERCISE 18B

NAME

INSTRUCTIONS:

1. Draw the objects shown above.
2. Dimension as shown using LINEAR and DIAMETER dimensioning.
3. NOTE: Dimension the Diameters first, then add center marks to the remaining circles.
   Draw the Linear dimensions last so you can snap to the endpoint of the center marks.
4. Save as: **EX-18B and plot.**

18-8

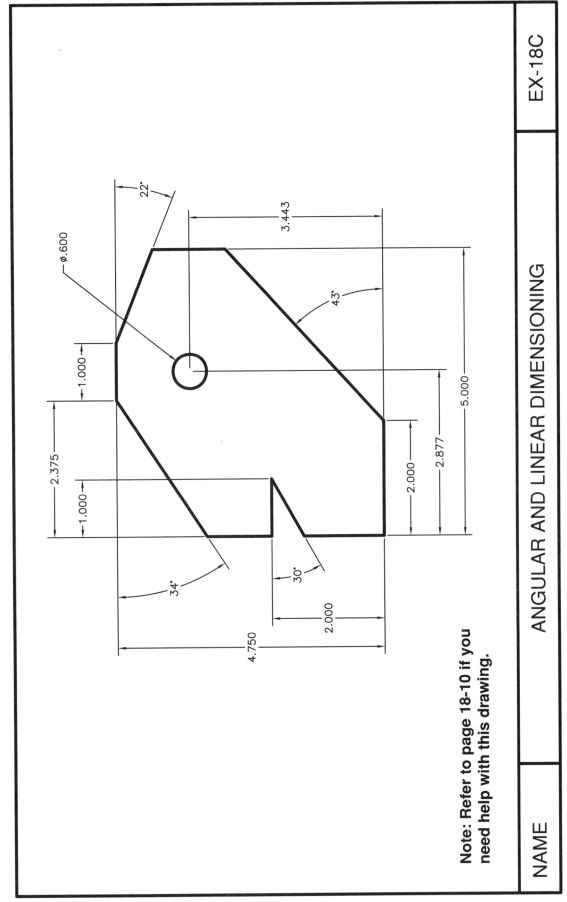

**Note: Refer to page 18-10 if you
need help with this drawing.**

ANGULAR AND LINEAR DIMENSIONING

EXERCISE 18C

| NAME | | |

INSTRUCTIONS:

1. Draw the object above. Consider using OFFSET instead of coordinate input.
2. Dimension as shown using LINEAR and ANGULAR dimensioning.
3. Dimension Style = Class style   Layer = DIM
4. Save as: **EX-18C and plot.**

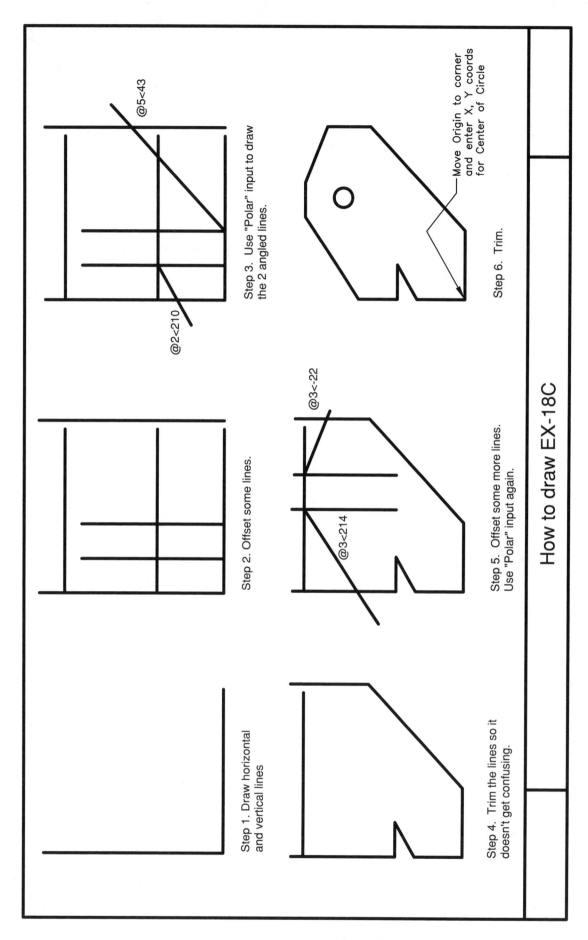

Step 1. Draw horizontal and vertical lines

Step 2. Offset some lines.

Step 3. Use "Polar" input to draw the 2 angled lines.

@5<43

@2<210

Step 4. Trim the lines so it doesn't get confusing.

Step 5. Offset some more lines. Use "Polar" input again.

@3<-22

@3<214

Step 6. Trim.

Move Origin to corner and enter X, Y coords for Center of Circle

How to draw EX-18C

EX-18C Helper

LEARNING OBJECTIVES

After completing this lesson, you will be able to:

1. Dimension objects that are on an angle.
2. Draw a Leader.
3. Draw a line with an arrow (no text)
4. Add symbols such as: diameter, plus or minus and degree to text.
5. Pre-assign a prefix or suffix to a dimension.

LESSON 19

ALIGNED DIMENSIONING

The **ALIGNED** dimension command aligns the dimension with the angle of the object that you are dimensioning. The process is the same as Linear dimensioning. It requires two extension line origins and placement of text location. (Example below)

1. Select the **ALIGNED** command using one of the following:

   TYPE = DIMALIGNED or DIMALI or DAL
   PULLDOWN = DIMENSION / ALIGNED
   TOOLBAR = DIMENSION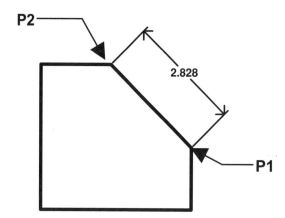

   Command: _dimaligned
2. Specify first extension line origin or <select object>: *select the first extension line origin (P1)*
3. Specify second extension line origin: *select the second extension line origin (P2)*

4. Specify dimension line location or [Mtext/Text/Angle]: *place dimension text location*
   Dimension text = *the dimension value will appear here*

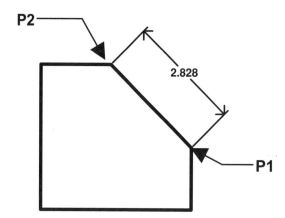

LEADER

The **LEADER** command is primarily used to add a **note** to an object. A Leader's appearance is very similar to a Radial dimension **but** a Leader **should not** to be used for Radial dimensioning. A Leader allows you to type a note at the end of the hook. (Note: Leader and Qleader are the same command)

If associative dimensioning is turned on with DIMASSOC, the leader start point can be associated with a location on an object. If the object is relocated, the arrowhead remains attached to the object and the leader line stretches, but the text or feature control frame remains in place.

1. Select the **LEADER** command using one of the following.

   TYPE = LE or QLEADER
   PULLDOWN = DIMENSION / LEADER
   TOOLBAR = DIMENSION ⬚

   See "Settings" below

   Command: _qleader
2. Specify first leader point, or [Settings]<Settings>: *select the location for the arrowhead (P1)*
3. Specify next point: *select next location (P2)*
4. Specify next point: *select another point or press <enter>*
5. Specify text <u>width</u> <.000>: *press <enter> (NOTE: This is not asking for Height)*
6. Enter first line of annotation text <Mtext>: *type your desired text here*
7. Enter next line of annotation text: *type more text or press <enter> to stop*

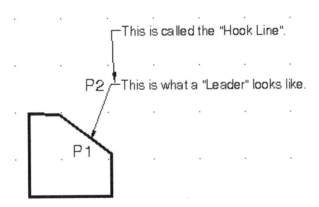

This is called the "Hook Line".

P2 ─ This is what a "Leader" looks like.

P1

HOOK LINE
The hook line is automatically added to the last line segment (P2) if the leader line is 15 degrees or more from horizontal. The length of the Hook Line is controlled by the arrow length setting in the dimension style.

SETTINGS:
If you would like to make changes to the appearance of the Leader, select **"Settings"** before you place the first point (arrowhead location). Selecting **"Settings"** will display a dialog box with many options. (These options are discussed in the Advanced Workbook.)

LINE WITH ARROW ONLY (NO TEXT)

Occasionally you will need to draw a line with an arrow on the end. But you do not want text at the other end. Just a line with an arrow pointing at something within your drawing.

To accomplish this you must tell AutoCAD that you do not want text. (AutoCAD calls this "annotation") You must change the "annotation type" setting to "None".

CHANGING THE SETTING FOR ANNOTATION TYPE

1. Select the **LEADER** command using one of the following.

   TYPE = LE or QLEADER
   PULLDOWN = DIMENSION / LEADER
   TOOLBAR = DIMENSION

   Command: _qleader
2. Specify first leader point, or [Settings]<Settings>: *select the "Settings" option*

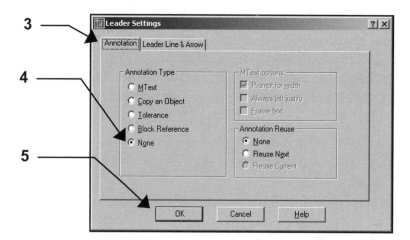

3. Select the "**Annotation**" tab.
4. Select "**None**"
5. Select the "**OK**" button.

Now continue drawing the arrow and line.

6. Specify first leader point, or [Settings]<Settings>: *select the location for the arrowhead (P1)*
7. Specify next point: *select the location for the other end of the line (P2)*
8. Specify next point: *press <enter> to stop*

P1 ◄─────────────────────── P2

Note: You will have to change Annotation Type back to "Mtext" if you wish to create a leader with text.

SPECIAL TEXT CHARACTERS

Characters such as the *degree symbol, diameter symbol* and the *plus / minus symbols* are created by typing **%%** and then the appropriate "code" letter.
For example: entering 350**%%D** will create: **350°**. The **"D"** is the **"code"** letter.

SYMBOL		CODE
∅	Diameter	%%C
°	Degree	%%D
±	Plus / Minus	%%P

SINGLE LINE TEXT

If you are using "Single Line Text", type the code in the sentence. While you are typing, the code will appear, in the sentence, on the command line. But when you are finished typing, and press <enter>, the symbol will appear.

MULTILINE TEXT

The code displays the same as single line text if you type your text in the text editor. But multiline text offers another method for inserting special character "symbols".

1. While you are typing in the in-place text editor, instead of typing the code, right click and the menu will appear.

2. Select "Symbol".

3. Select the symbol desired and the code or the actual symbol will automatically appear in the sentence.

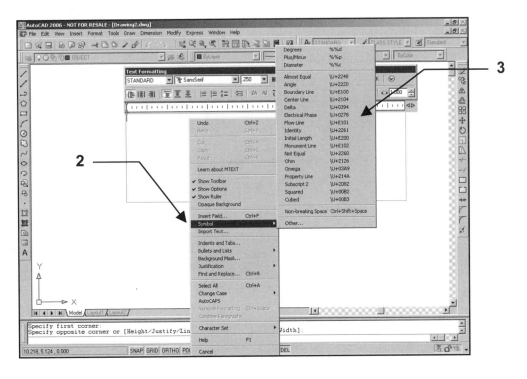

PREFIX and SUFFIX

The **PREFIX** (before) and the **SUFFIX** (after) allows you to preset text to be inserted automatically as you dimension.

Primary Units Tab

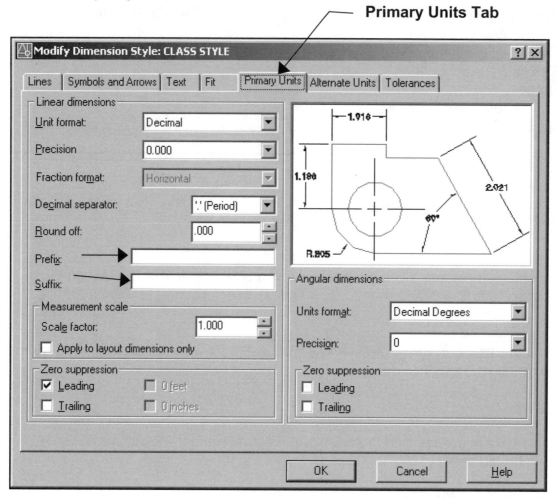

Examples:

If you wanted all of your dimensions to end with **Ref**, you would type **<space>Ref** in the suffix box.

If you wanted all of your dimensions to begin with **2 places**, you would type **2 places** in the prefix box.

Note: If you enter text in the Prefix box when drawing Radial dimensions, the **"R"** for radius or the **diameter symbol**, <u>will not</u> be automatically drawn. You must also add the symbol code to the prefix box.

Example: If you would like a diameter dimension to have 2X as a Prefix, you must type the following, in the Prefix box: **2X <space> %%C**.

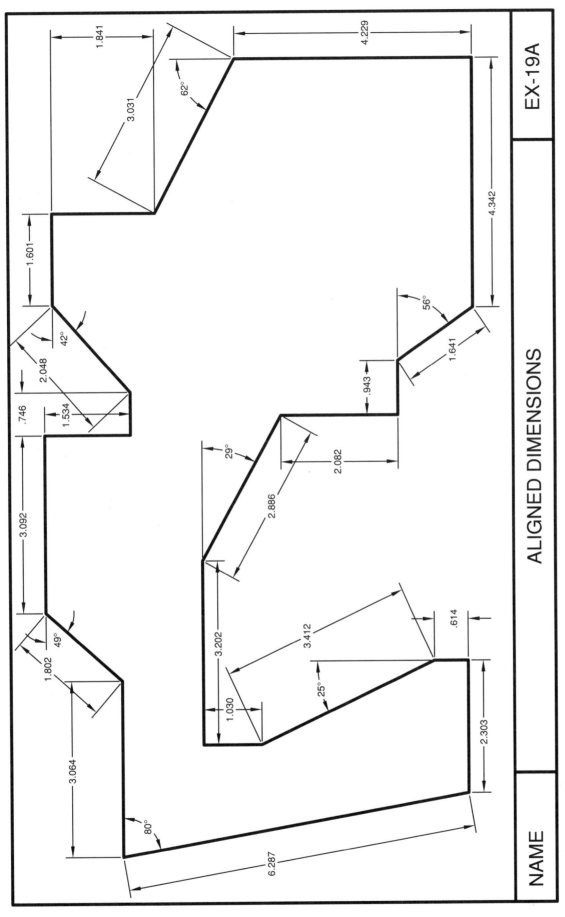

EXERCISE 19A

EX-19A

ALIGNED DIMENSIONS

NAME

INSTRUCTIONS:

1. Draw the object shown above. You will get to practice "Polar Coordinates" or "Dynamic Input".

2. Dimension using Aligned, Linear and Angular.

3. Use Dimension Style "Class Style".

4. Save as **EX19A and Plot**.

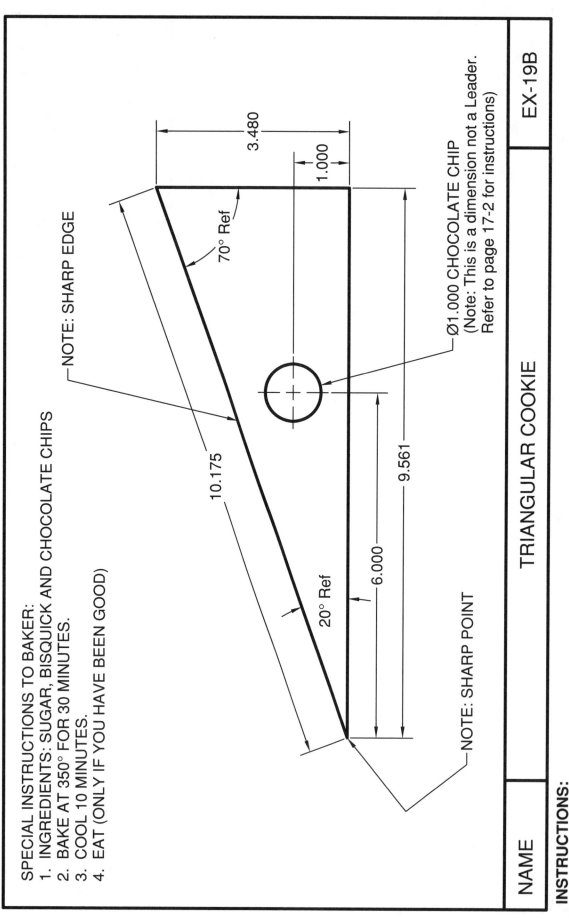

SPECIAL INSTRUCTIONS TO BAKER:
1. INGREDIENTS: SUGAR, BISQUICK AND CHOCOLATE CHIPS
2. BAKE AT 350° FOR 30 MINUTES.
3. COOL 10 MINUTES.
4. EAT (ONLY IF YOU HAVE BEEN GOOD)

3.480

1.000

70° Ref

NOTE: SHARP EDGE

Ø1.000 CHOCOLATE CHIP
(Note: This is a dimension not a Leader.
Refer to page 17-2 for instructions)

10.175

9.561

20° Ref

6.000

NOTE: SHARP POINT

NAME	TRIANGULAR COOKIE	EX-19B

EXERCISE 19B

INSTRUCTIONS:
1. Draw the object above.
2. Dimension as shown using:
   LINEAR, ANGULAR, RADIAL, ALIGNED and LEADER dimensioning.
3. Use Dimension Style "Class Style".
4. For the Note "Special instructions to Baker" use: Text Ht. = .200   Layer = TXT-LIT
5. Save as: **EX-19B and plot.**

LEARNING OBJECTIVES

After completing this lesson, you will be able to:

1. Use Multiple Automatic dimensioning.
2. Edit Multiple dimensions.

Sorry LT users, the information in this lesson is not available within the LT version.

LESSON 20

QUICK DIMENSION

Sorry LT users, this command is not available to you. Skip to Lesson 21.

Quick Dimension creates multiple dimensions with one command. Quick Dimension can create Continuous, Staggered, Baseline, Ordinate, Radius and Diameter dimensions. *("Ordinate" will be discussed in the Advanced workbook)*
Qdim only works in modelspace. It is not trans-spatial. (Refer to page 27-10)

1. Select the **Quick Dimension** command using one of the following:

> **TYPE = QDIM**
> **PULLDOWN = Dimension/Quick Dimension**
> **TOOLBAR = DIMENSION**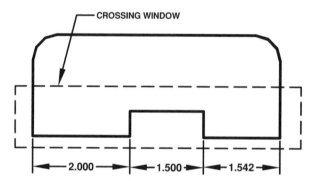
>
> Command: _qdim
> Associative dimension priority = Endpoint
> Select geometry to dimension

CONTINUOUS

2. Select the objects to be dimensioned with a crossing window or pick each object
3. Press <enter> to stop
4. Specify dimension line position, or
[Continuous/Staggered/Baseline/Ordinate/Radius/Diameter/datumPoint/Edit/settings]
<Continuous>: *Select "C" <enter> for Continuous.*
5. Select the location of the dimension line.

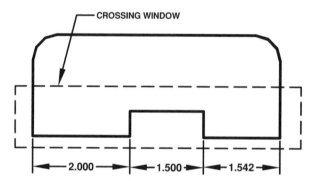

BASELINE

2. Select the objects to be dimensioned with a crossing window or pick each object.
3. Press <enter> to stop
4. Specify dimension line position, or
[Continuous/Staggered/Baseline/Ordinate/Radius/Diameter/datumPoint/Edit/settings]
<Continuous>: *Select "B" <enter> for Baseline.*
5. Select the location of the dimension line.

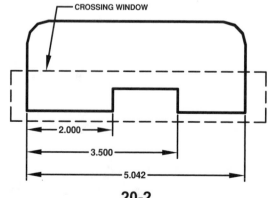

20-2

STAGGERED

2. Select the objects to be dimensioned with a crossing window or pick each object.
3. Press <enter> to stop
4. Specify dimension line position, or
[Continuous/Staggered/Baseline/Ordinate/Radius/Diameter/datumPoint/Edit/settings]
<Continuous>: *Select "S" <enter> for Staggered.*
5. Select the location of the dimension line.

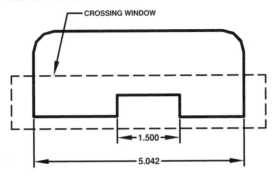

DIAMETER

2. Select the objects to be dimensioned with a crossing window or pick each object
   (Qdim will automatically filter out any linear dims)
3. Press <enter> to stop.
4. Specify dimension line position, or
[Continuous/Staggered/Baseline/Ordinate/Radius/Diameter/datumPoint/Edit/settings]
<Continuous>: *Select "D" <enter> for Diameter*
5. Select the location of the dimension line.
(Dimension line length is determined by the "Baseline Spacing" setting)

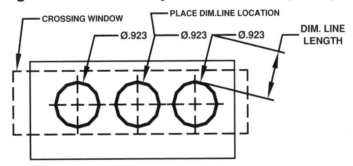

RADIUS

2. Select the objects to be dimensioned with a crossing window or pick each object.
   (Quick Dimension will automatically filter out any linear dimensions)
3. Press <enter> to stop selecting objects
4. Specify dimension line position, or
[Continuous/Staggered/Baseline/Ordinate/Radius/Diameter/datumPoint/Edit/settings]
<Continuous>: *Select "R" <enter> for Radius*
The dimensions are automatically placed, you do not select the location for the
dimension line. Dimension line length is determined by the "Baseline Spacing" setting)

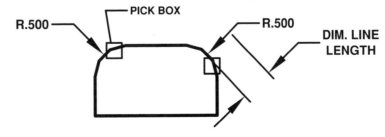

EDITING MULTIPLE DIMENSIONS

You can edit existing multiple dimensions using the QDIM / Edit command. The Qdim, edit command will edit all multiple dimensions, no matter whether they were created originally with Qdim or not. All multiple: linear, baseline and continue dimensions respond to this editing command.

1. Select the **QDIM** command.
2. Select the dimensions to edit.

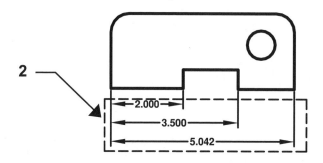

3. Press <enter> to stop selecting.
4. Select "E" for edit.

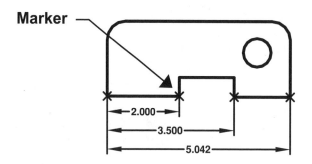

Small *markers* will appear at the extension line origins.
These markers are called **"dimension points"**.
You can **Add** or **Remove** dimensions by selecting these markers.

*See **ADD** and **REMOVE** on the next page.*

TO **REMOVE** A DIMENSION

AutoCAD assumes that you want to Remove dimensions, so this is the default setting.

5. Indicate dimension point to remove, or [Add/eXit] <eXit>: **Click on the extension line markers you want to remove**

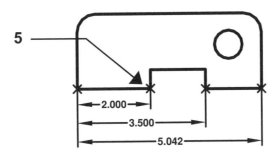

6. Press "X" to stop selecting.
7. Move the cursor to place the remaining dimensions again.

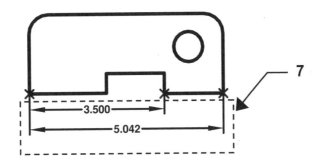

TO **ADD** A DIMENSION

8. Select "A" <enter> for Add
9. Using Object Snap, select the extension line origins of the dimension you want to Add.
10. Press "X" to stop selecting.
11. Replace the dimension line location.

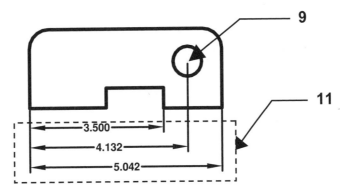

Why would I bother using ADD and REMOVE when I can just use Erase or add a dimension? The Qdim ADD and REMOVE options automatically respace the dimensions. You will not have to stretch or move the remaining dimensions.

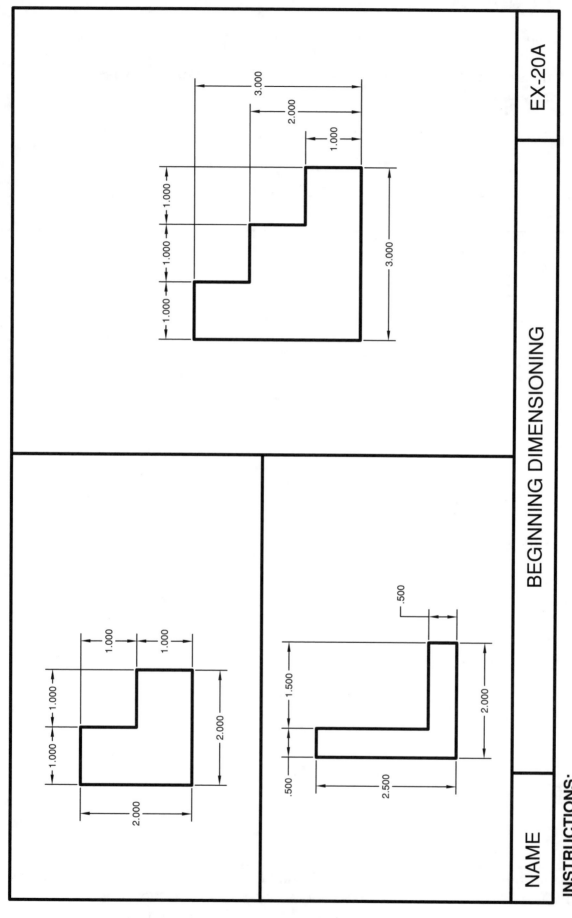

EXERCISE 20A

EX-20A

BEGINNING DIMENSIONING

NAME

INSTRUCTIONS:

1. Open EX-16A
2. Erase all the dimensions
3. Dimension again using QDIM command.
4. Save as: **EX-20A and plot.**

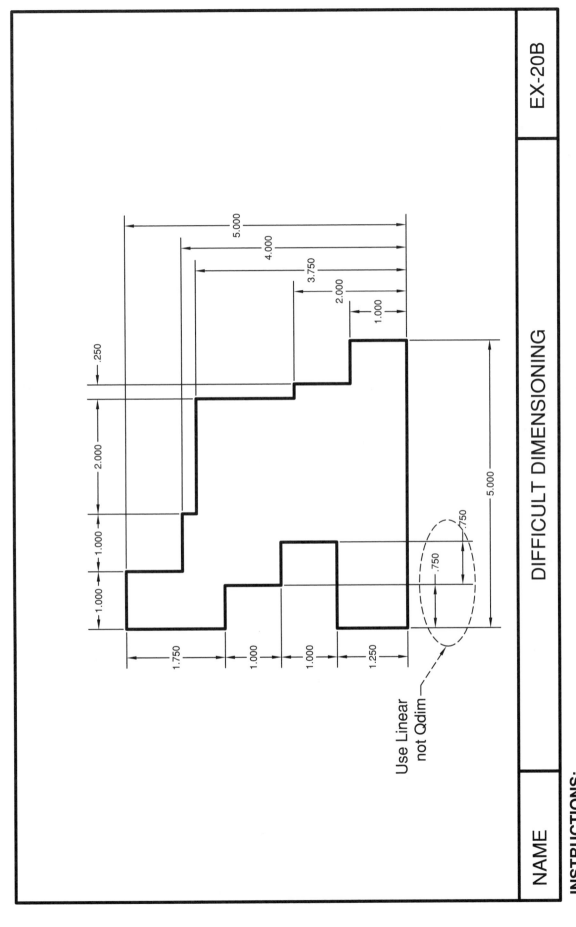

EXERCISE 20B

NAME	DIFFICULT DIMENSIONING	EX-20B

INSTRUCTIONS:

1. Open EX-16B.
2. Erase all of the Dimensions.
3. Dimension again using QDIM.
4. Save as: **EX-20B and plot.**

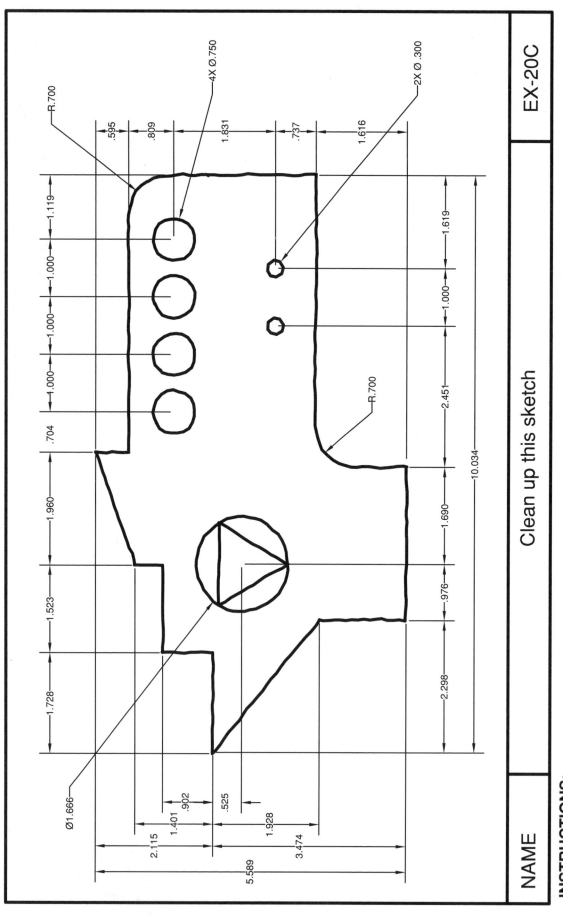

EXERCISE 20C

| NAME | Clean up this sketch | EX-20C |

INSTRUCTIONS:

Many times, on the job, you will be handed a sketch like the one shown above. You will be expected to create a clean and accurate drawing from this sketch. After drawing the objects accurately, try out the new Quick dimension command. You will find that you can use Quick dimension for some but you must use Linear, Continue and Baseline for others. You decide which is best.

Save as: **EX-20C**

LEARNING OBJECTIVES

After completing this lesson, you will be able to:

1. Change an object's properties to match the properties of another object.
2. Create a Revision Cloud
3. Select the Revision Cloud Style
4. Cover part of the drawing with a blank patch.

LESSON 21

MATCH PROPERTIES

Match Properties is used to "paint" the properties of one object to another. This is a simple and useful command. You first select the object that has the desired properties (the source object) and then select the object you want to "paint" the properties to (destination object).

Only one "source object" can be selected but its properties can be painted to any number of "destination objects".

1. Select the Match Properties command using one of the following:

   TYPE = MATCHPROP or MA
   PULLDOWN = MODIFY / MATCH PROPERTIES
   TOOLBAR = STANDARD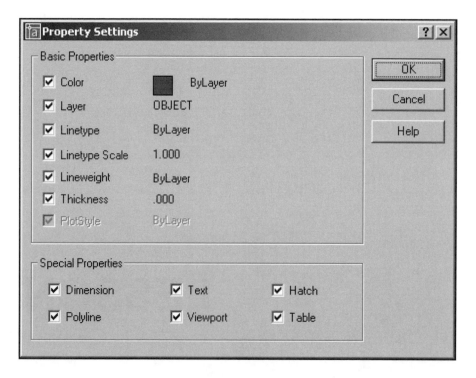

   Command: matchprop

2. Select source object: ***select the object with the desired properties to match***

3. Select destination object(s) or [Settings]: ***select the object(s) you want to receive the matching properties.***

4. Select destination object(s) or [Settings]: ***select more objects or <enter> to stop.***

Note: If you do not want to match all of the properties, right click and select "Settings" from the short cut menu, before selecting the destination object. Uncheck all the properties you do not want to match and select the OK button. Then select the destination object(s).

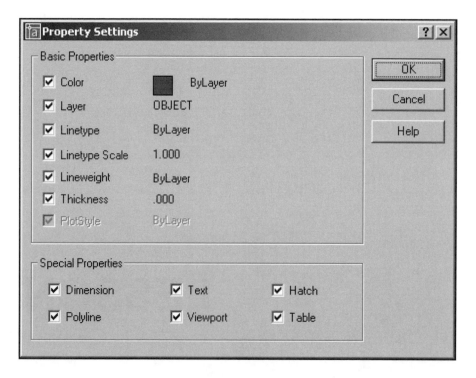

CREATING A REVISION CLOUD

When you make a revision to a drawing it is sometimes helpful to highlight the revision for someone viewing the drawing. A common method to highlight the area is to draw a "Revision Cloud" around the revised area. This can be accomplished easily with the "Revision Cloud" command.

The Revision Cloud command creates a series of sequential arcs to form a cloud-shaped object. You set the minimum and maximum arc lengths. (Maximum arc length cannot exceed three times the minimum arc length. Example: Min = 1, Max can be 3 or less) If you set the minimum and maximum different lengths the arcs will vary in size and will display an irregular appearance.

Min & Max same length   **Min & Max different length**

To draw a Revision Cloud you specify the start point with a left click then drag the cursor to form the outline. AutoCAD automatically draws the arcs. When the cursor gets very close to the start point, AutoCAD snaps the last arc to the first arc and closes the shape.

1. Select the Revision Cloud command using one of the following:

   TYPE = REVCLOUD
   PULLDOWN = DRAW / REVISION CLOUD
   PULLDOWN (LT) = TOOLS / REVISION CLOUD
   TOOLBAR = DRAW

   Command: _revcloud
   Minimum arc length: .50   Maximum arc length: .50   Style: Normal

2. Specify start point or [Arc length/Object/Style] <Object>: *Select "Arc length"*

3. Specify minimum length of arc <.50>: *Specify the minimum arc length*

4. Specify maximum length of arc <.50>: *Specify the maximum arc length*

5. Specify start point or [Arc length/Object/Style] <Object>: *Place cursor at start location & left click.*

6. Guide crosshairs along cloud path...*Move the cursor to create the cloud outline.*

7. Revision cloud finished. *When the cursor approaches the start point, the cloud closes automatically.*

CONVERT A CLOSED OBJECT INTO A REV CLOUD

You can convert a closed object, such as a circle, ellipse, rectangle or closed polyline to a revision cloud. The original object is deleted when it is converted.
(If you want the original object to remain, in addition to the new rev cloud, set the variable "delobj" to "0". The default setting is "1".)

1. Draw a closed object such as a circle.

2. Select the Revision Cloud command using one of the following:

   TYPE = REVCLOUD
   PULLDOWN = DRAW / REVISION CLOUD
   PULLDOWN (LT) = TOOLS / REVISION CLOUD
   TOOLBAR = DRAW

   Command: _revcloud
   Minimum arc length: .50   Maximum arc length: .50   Style: Normal

3. Specify start point or [Arc length/Object/Style] <Object>: *Select "Arc length"*

4. Specify minimum length of arc <.50>: *Specify the minimum arc length*

5. Specify maximum length of arc <.50>: *Specify the maximum arc length*

6. Specify start point or [Arc length/Object/Style] <Object>: *Select "Object".*

7. Select object: *Select the object to convert*

8. Select object: Reverse direction [Yes/No] <No>: *Select Yes or No*
   Revision cloud finished.

REVERSE DIRECTION

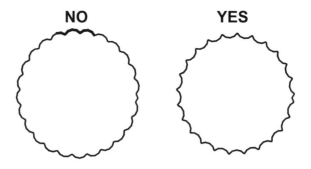

NO   YES

NOTE:
The Match Properties command will not match the arc length from the source cloud to the destination cloud.

REVISION CLOUD STYLE

You may select one of 2 styles for the Revision Cloud; **Normal** or **Calligraphy**.
Normal will draw the cloud with one line width.
Calligraphy will draw the cloud with variable line widths to appear as though you used a chiseled calligraphy pen.

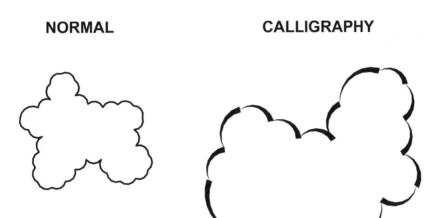

NORMAL   **CALLIGRAPHY**

1. Select the Revision Cloud command using one of the following:

    TYPE = REVCLOUD
    PULLDOWN = DRAW / REVISION CLOUD
    PULLDOWN (LT) = TOOLS / REVISION CLOUD
    TOOLBAR = DRAW

    Command: _revcloud
    Minimum arc length: .50   Maximum arc length: 1.00   Style: Normal

2. Specify start point or [Arc length/Object/Style] <Object>: *Select "Style"<enter>*

3. Select arc style [Normal/Calligraphy] <Calligraphy>:*Select "N or C"<enter>*

4. Specify start point or [Arc length/Object/Style] <Object>: *Select "Arc length"*

5. Specify minimum length of arc <.50>: *Specify the minimum arc length*

6. Specify maximum length of arc <1.00>: *Specify the maximum arc length*

7. Specify start point or [Object] <Object>: *Place cursor at start location & left click.*

8. Guide crosshairs along cloud path...*Move the cursor to create the cloud outline.*

9. Revision cloud finished. *When the cursor approaches the start point, the cloud closes automatically.*

WIPEOUT

The Wipeout command creates a blank area that covers existing objects. The area has a background that matches the background of the drawing area. This area is bounded by the wipeout frame, which you can turn on or off.

1. Select the Wipeout command using one of the following:

   TYPE = WIPEOUT
   PULLDOWN = DRAW / WIPEOUT
   TOOLBAR = NONE

2. Command: _wipeout Specify first point or [Frames/Polyline] <Polyline>: *specify the first point of the shape (P1)*
3. Specify next point: *specify the next point (P2)*
4. Specify next point or [Undo]: *specify the next point (P3)*
5. Specify next point or [Undo]: *specify the next point (P4)*
6. Specify next point or [Close/Undo]: *specify the next point or <enter> to close*

BEFORE WIPEOUT   **AFTER WIPEOUT**

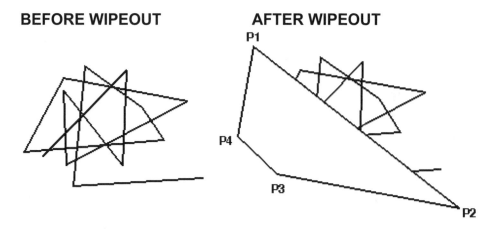

TURNING FRAMES ON OR OFF

1. Select the Wipeout command.
2. Select the "Frames" option.
3. Enter ON or OFF.

ON   **OFF**

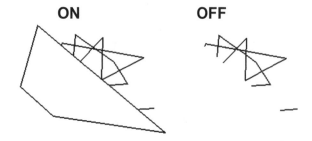

Note: If you want to move the objects and the wipeout area, move them together not separately. If you do move the objects and the wipeout, the objects under the wipeout area may reappear. Select **View / Regen** and they will disappear again.

EXERCISE 21A

INSTRUCTIONS:

1. Open **EX 5F.dwg**.
2. Using **Match Properties,** change the properties of the Polygon to the same properties as the Donut.
3. Using Wipeout, block out the bottom donut approximately as shown.
4. **Save** this drawing as: **EX21A and Plot.**

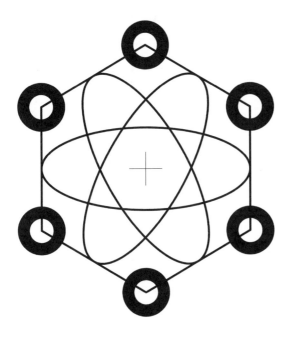

BEFORE

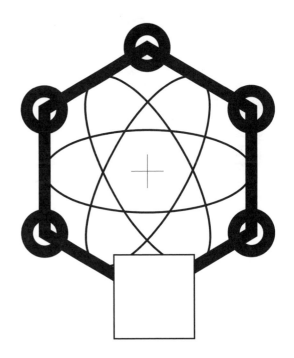

AFTER

EXERCISE 21B

Draw (2) Ø3" Circles and convert to Revision Clouds.
Set Min. and Max. Arc length to .500.
#1: Reverse = No
#2: Reverse = Yes

No

#1

Yes

#2

EX-21B

REVISION CLOUD

Draw 3 Revision Clouds
-#1: with Min. and Max. length set to .500
-#2: with Min. set to .500 & Max. set to 1.500.
-#3: same as #2 but change style to calligraphy
Make any shape you like.

#1

#2

#3

NAME

INSTRUCTIONS:
1. Follow the instructions above.
2. Save as: **EX-21B and plot.**

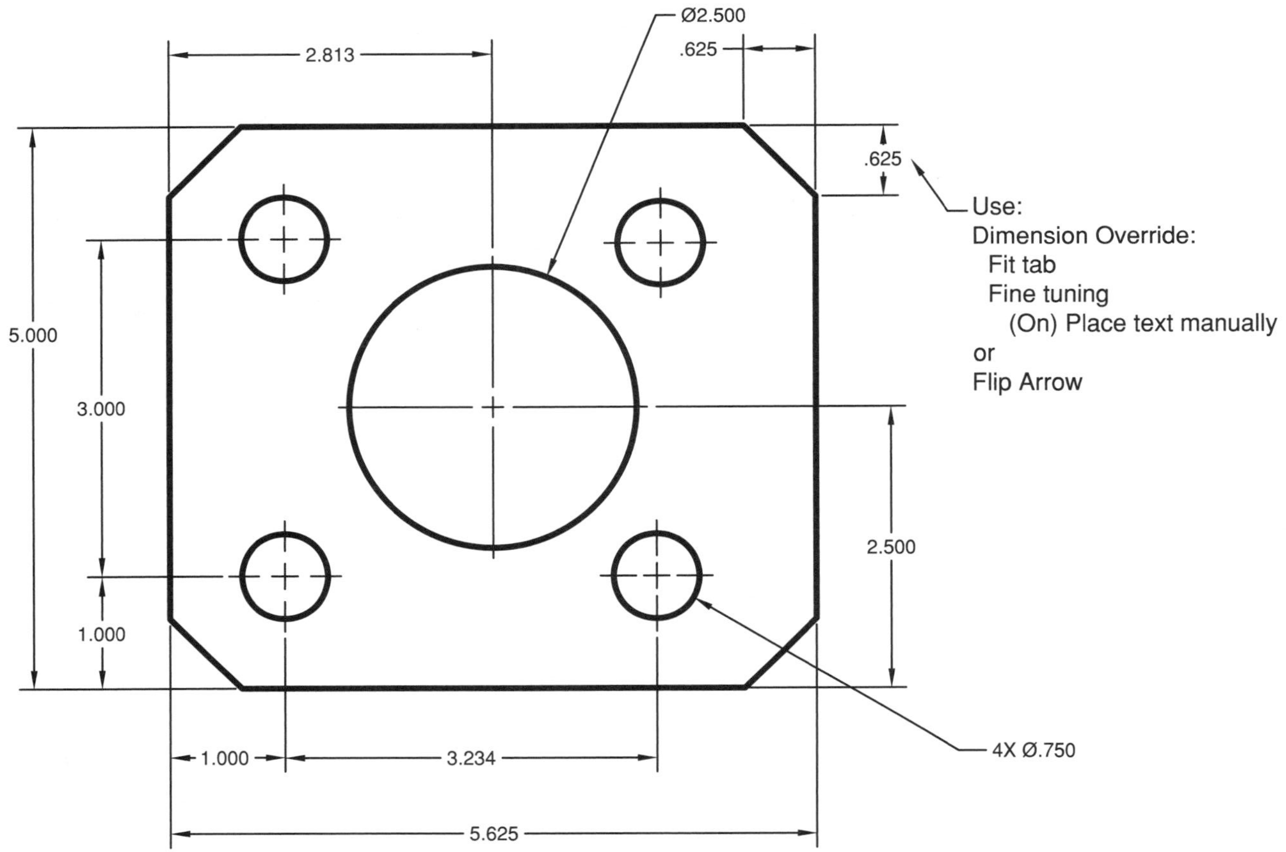

Use:
Dimension Override:
Fit tab
Fine tuning
(On) Place text manually
or
Flip Arrow

Ø2.500

.625

.625

2.813

4X Ø.750

2.500

3.234

5.625

1.000

3.000

1.000

5.000

EXERCISE 21C

| NAME | | BASE PLATE | EX-21C |

INSTRUCTIONS:

The drawing above should give you additional practice with drawing objects and dimensioning.

1. Draw the objects above and dimension.

2. Remember when drawing a RECTANGLE you may chamfer the corners.
   (Refer to Lesson 3)

3. Save as: **EX-21C and plot.**

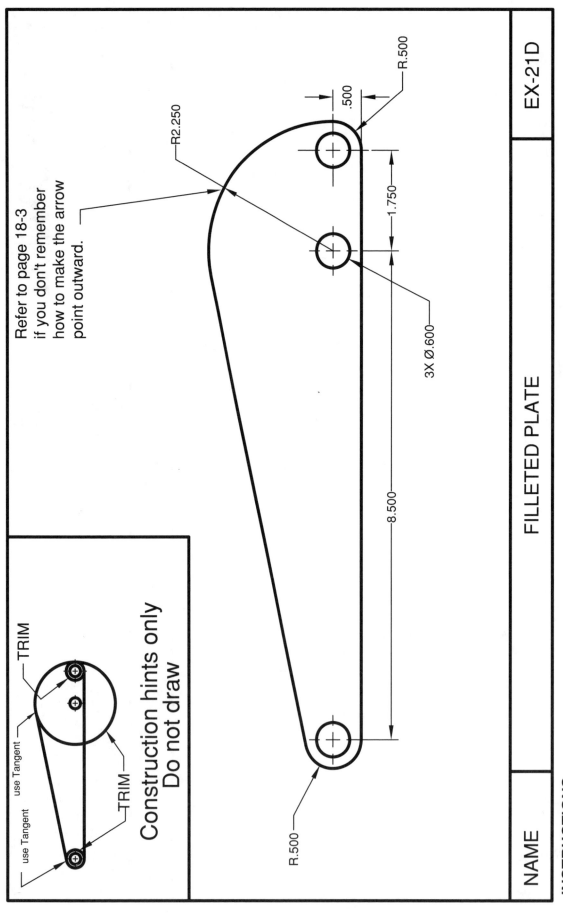

Refer to page 18-3
if you don't remember
how to make the arrow
point outward.

R2.250

.500

R.500

1.750

3X Ø.600

8.500

R.500

TRIM

use Tangent

use Tangent

TRIM

Construction hints only
Do not draw

EXERCISE 21D

EX-21D

FILLETED PLATE

NAME

INSTRUCTIONS:

This drawing should give you additional practice with Radial dimensioning and Leader.

1. Draw and dimension the FILLETED PLATE above using:
   Lines, Offset, Copy, and Trim.
2. Dimension as shown.
3. Save as: **EX-21D and plot.**

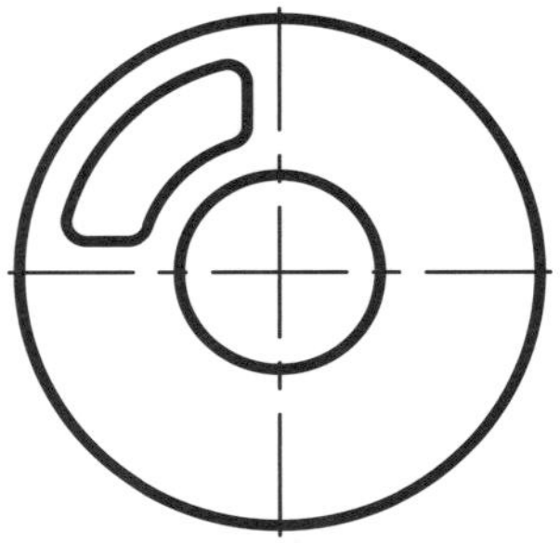

EXERCISE 21E

Ø6.500

1.480

R1.250

Use "Flip Arrow" see page 18-3

.400

.200

R1.750

Use "Override" to
change settings
see pages 17-5 & 18-3

R.250

R2.750

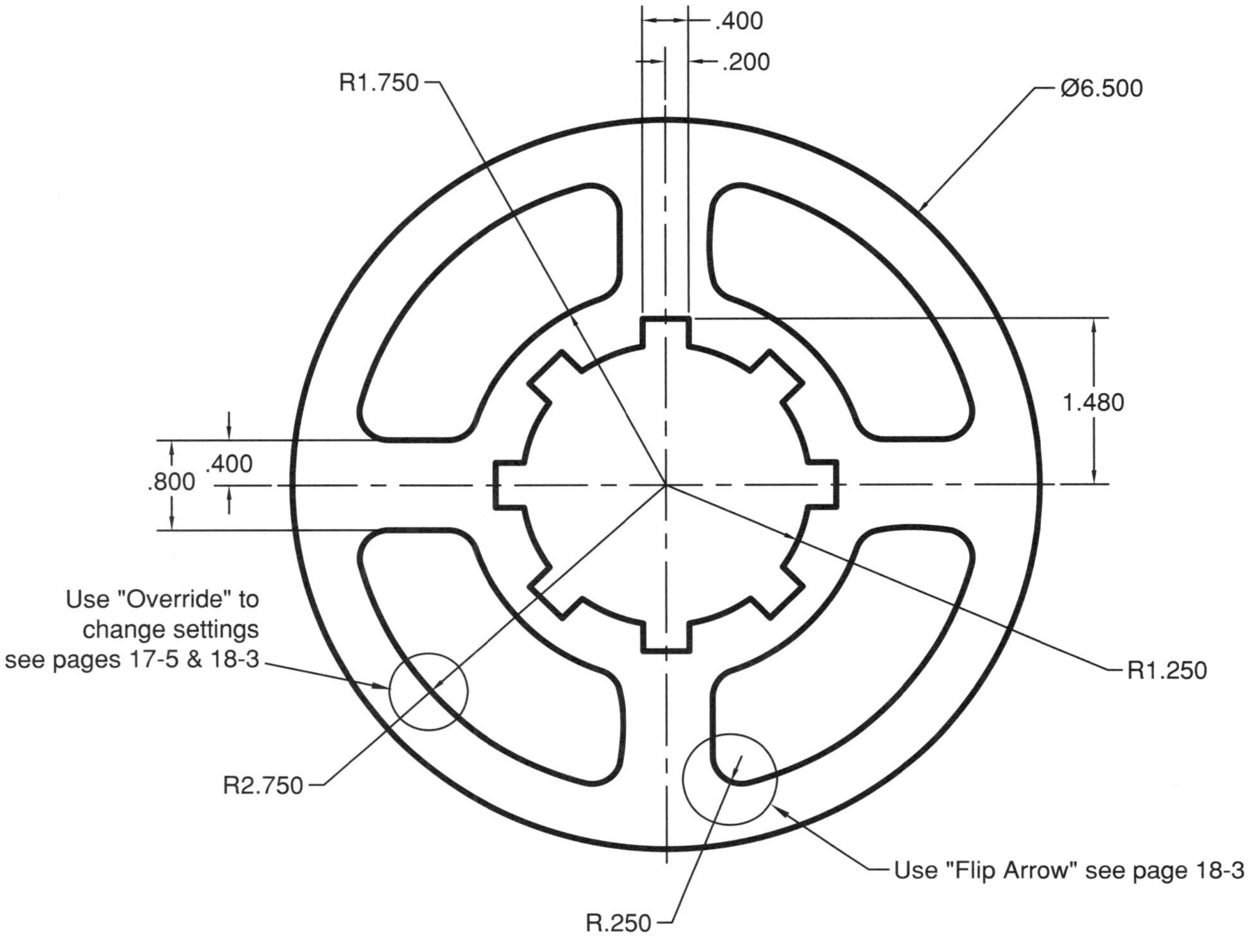

.400

.800

	EX-21E

CIRCULAR PATTERN

Construction
Hints

Draw 1 segment then
use polar array.

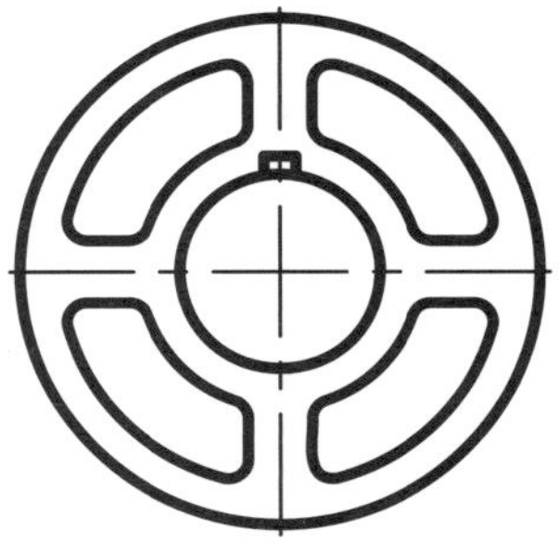

Draw 1 slot then use
polar array then trim.

NAME	

INSTRUCTIONS:
This drawing will give you additional practice with the polar array command.
1. Draw and dimension the FILLETED PLATE above using:
   Lines, Circles, Offset, Array, Fillet and Trim.
2. Dimension as shown.
3. Save as: **EX-21E and plot.**

NOTES:

LEARNING OBJECTIVES

After completing this lesson, you will be able to:

1. Draw an Arc using 10 different methods.
2. Dimensioning an Arc.

LESSON 22

ARC

TYPING = A <enter>
PULLDOWNS = DRAW / ARC
TOOLBARS =DRAW

There are 10 ways to draw an ARC in AutoCAD.   Not all of the ARCS options are easy to create so you may find it is often easier to **trim a Circle** or use the **Fillet** command.

On the job, you will probably only use 2 of these methods. Which 2 depends on the application.

An **ARC** is a segment of a circle and must be less than 360 degrees.

Most ARCS are drawn counter-clockwise but you will notice in the examples on the following pages that some may be drawn clockwise by entering a negative input.

Examples of each of the ARC options are shown in EXERCISES 22A through 22C. Also, refer to the "Help" menu for additional examples of the use of the Arc command.

DIMENSIONING ARC LENGTHS

You may dimension the distance along an Arc. This is known as the Arc length.
Arc length is an associative dimension.

Example:

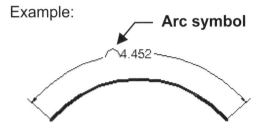

Arc symbol

Select the Arc length dimension command using one of the following:

TYPING = dimarc <enter>
PULLDOWNS = Dimension / Arc length
TOOLBARS =Dimension

Command: _dimarc
Select arc or polyline arc segment: *select the Arc*
Specify arc length dimension location, or [Mtext/Text/Angle/Partial/Leader]: *place the*
dimension line and text location
Dimension text = *dimension value will be shown here*

To differentiate the Arc length dimensions from Linear or Angular dimensions, arc length
dimensions display an arc (⌒) symbol by default. (Also called a "hat" or "cap")

The arc symbol may be displayed either <u>above,</u> or <u>preceding</u> the dimension text.
You may also choose not to display the arc symbol.

Example:

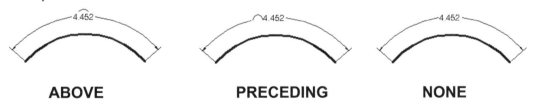

ABOVE   **PRECEDING**   **NONE**

Specify the placement of the arc symbol in the Dimension Style / Symbols and Arrows
tab or you may edit its position using the Properties Palette.

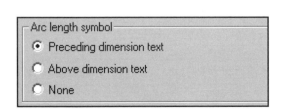

The extension lines of an Arc length dimension are displayed as <u>radial</u> if the included angle is <u>greater than 90 degrees</u>.

Example:

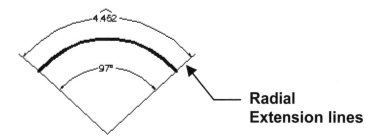

Radial
Extension lines

The extension lines of an Arc length dimension are displayed as <u>orthogonal</u> if the included angle is <u>less than 90 degrees</u>.

Example:

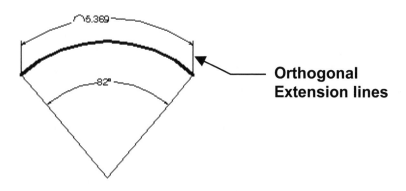

Orthogonal
Extension lines

DIMENSIONING A LARGE CURVE

When dimensioning an arc the dimension line should pass through the center of the arc. However, for large curves, the center of the arc could be very far away. Even off the sheet.

When the true center location cannot be displayed you can create a "Jogged" radius dimension.

Example:

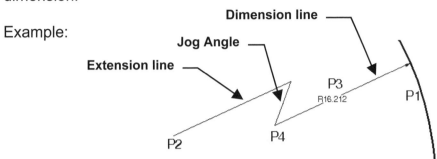

You can specify the jog angle in the Dimension Style / Symbols and Arrows tab.

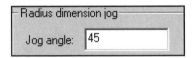

1. Select the Jogged radius dimension using one of the following:

TYPING = dimjogged <enter>
PULLDOWNS = Dimension / jogged
TOOLBARS = Dimension [icon]

2. Select arc or circle: *select the large arc or circle (P1 anywhere on arc)*

3. Specify center location override: *move the cursor and left click to specify the new center location (P2).*

   Dimension text = (actual radius will be displayed here)

4. Specify dimension line location or [Mtext/Text/Angle]: *move the cursor and left click to specify the location for the dimension text (P3).*

5. Specify jog location: *move the cursor and left click to specify the location for the jog (P4).*

Options:

Mtext: Displays the In-Place Text Editor, which you can use to edit the dimension text.

Text: You may customize the dimension text on the command line. The actual dimension is displayed in brackets < >.

Angle: Changes the angle of the dimension text.

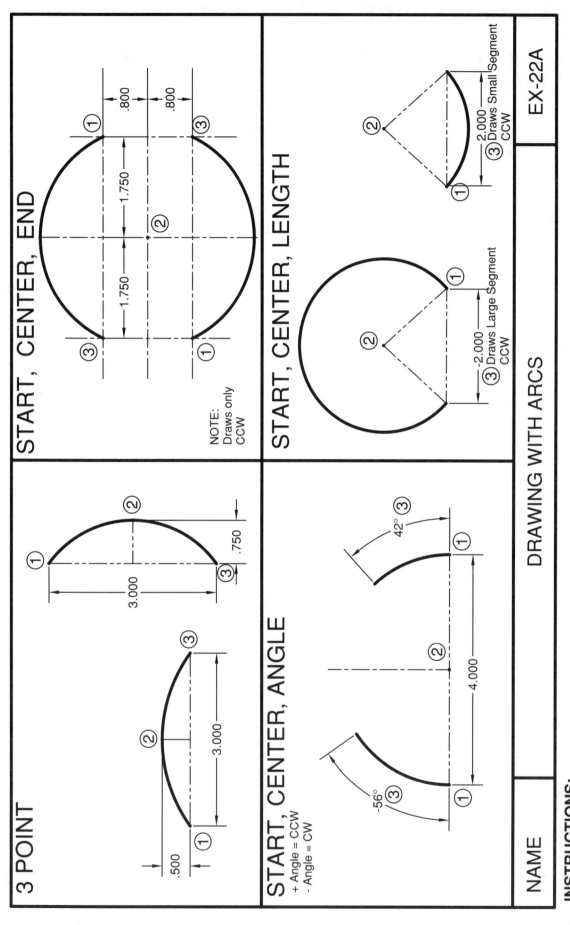

EXERCISE 22A

3 POINT

START, CENTER, END

NOTE:
Draws only
CCW

START, CENTER, ANGLE

+ Angle = CCW
- Angle = CW

START, CENTER, LENGTH

③ Draws Large Segment CCW

③ Draws Small Segment CCW

EX-22A

DRAWING WITH ARCS

NAME

INSTRUCTIONS:

1. Draw the ARCS above.
2. Select the method shown from the DRAW / ARC pulldown menu.
Note: Yours may not appear exactly as the example...that's OK.
3. Save as: **EX-22A.**

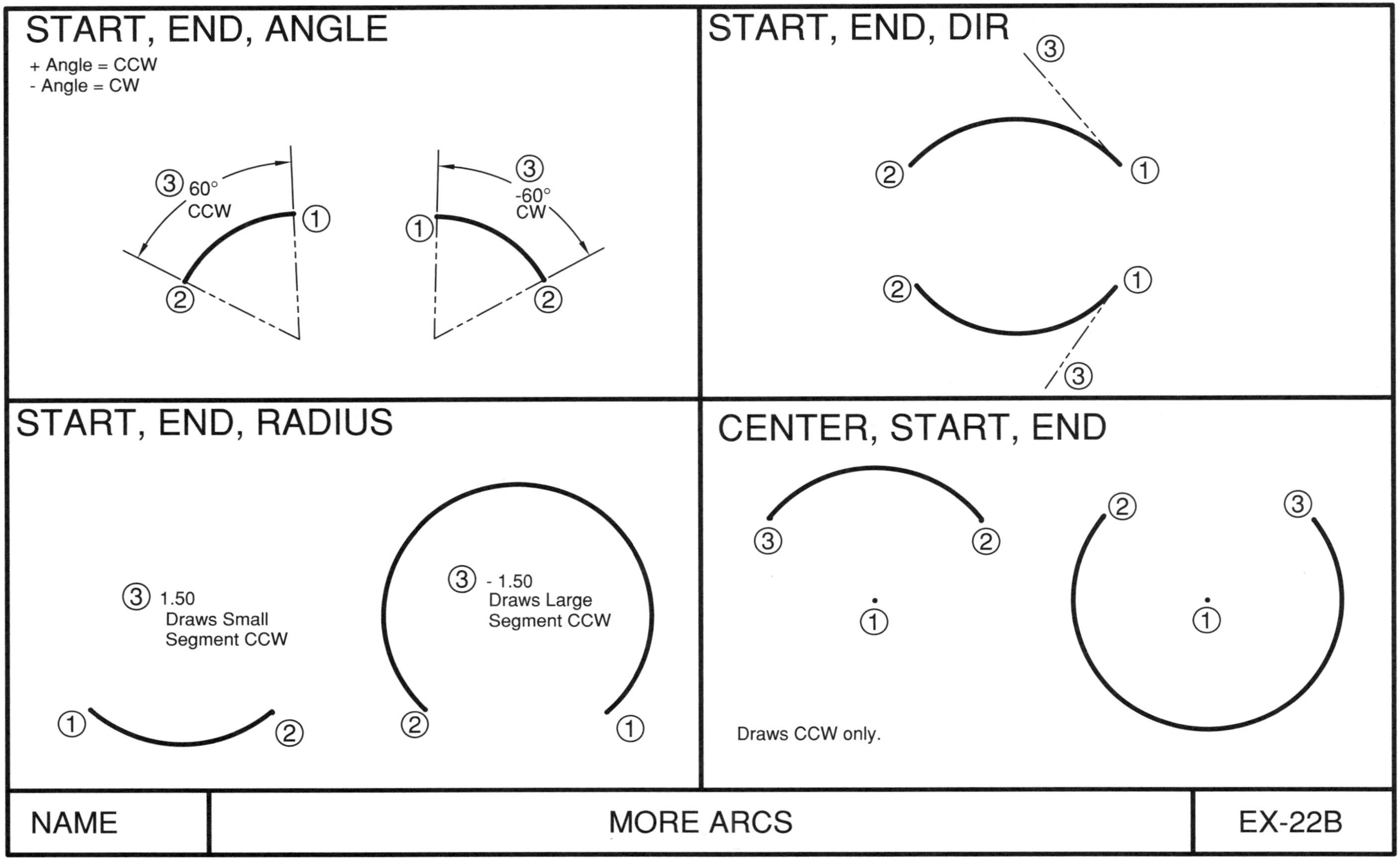

START, END, ANGLE

+ Angle = CCW
- Angle = CW

③ -60°
CW
②
①

③ 60°
CCW
①
②

START, END, DIR

①
③
②

③
①
②

CENTER, START, END

②
①
③

②
①
③

Draws CCW only.

START, END, RADIUS

③ 1.50
Draws Small
Segment CCW
②
①

③ -1.50
Draws Large
Segment CCW
①
②

MORE ARCS

NAME

EX-22B

EXERCISE 22B

INSTRUCTIONS:

1. Draw the ARCS above. Select the method shown.
2. Select the method shown from the DRAW / ARC pulldown menu.
Note: Yours may not appear exactly as the example...that's OK.
3. Save as: **EX-22B.**

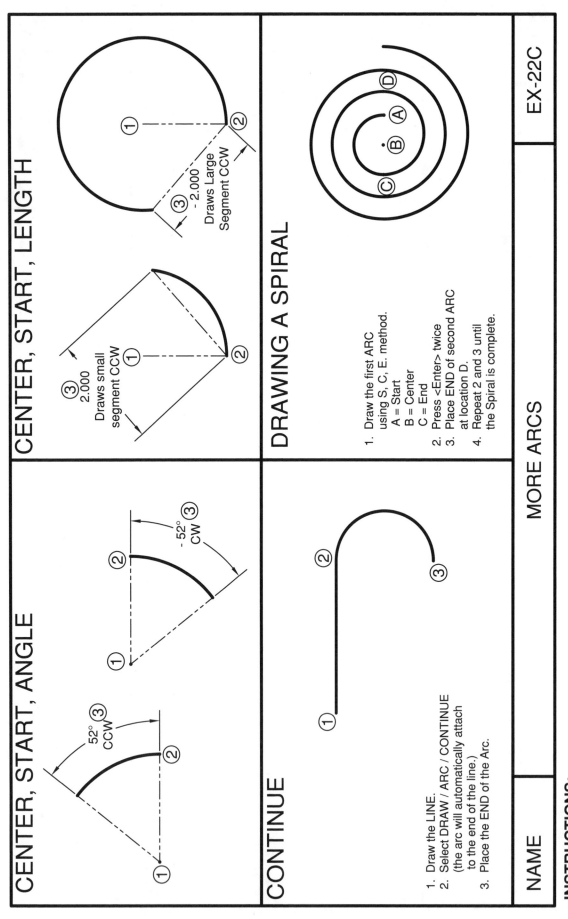

CENTER, START, ANGLE

52° ③ CCW

① ② ③

-52° CW

CENTER, START, LENGTH

-2.000
Draws Large Segment CCW

① ② ③

2.000
Draws small segment CCW

① ② ③

CONTINUE

① ② ③

1. Draw the LINE.
2. Select DRAW / ARC / CONTINUE (the arc will automatically attach to the end of the line.)
3. Place the END of the Arc.

DRAWING A SPIRAL

Ⓐ Ⓑ Ⓒ Ⓓ

1. Draw the first ARC using S, C, E. method.
   A = Start
   B = Center
   C = End
2. Press <Enter> twice
3. Place END of second ARC at location D.
4. Repeat 2 and 3 until the Spiral is complete.

NAME	MORE ARCS	EX-22C

EXERCISE 22C

INSTRUCTIONS:
1. Draw the ARCS above. Select the method shown.
2. Select the method shown from the DRAW / ARC pulldown menu.
Note: Yours may not appear exactly as the example...that's OK.
3. Save as: **EX-22C.**

22-8

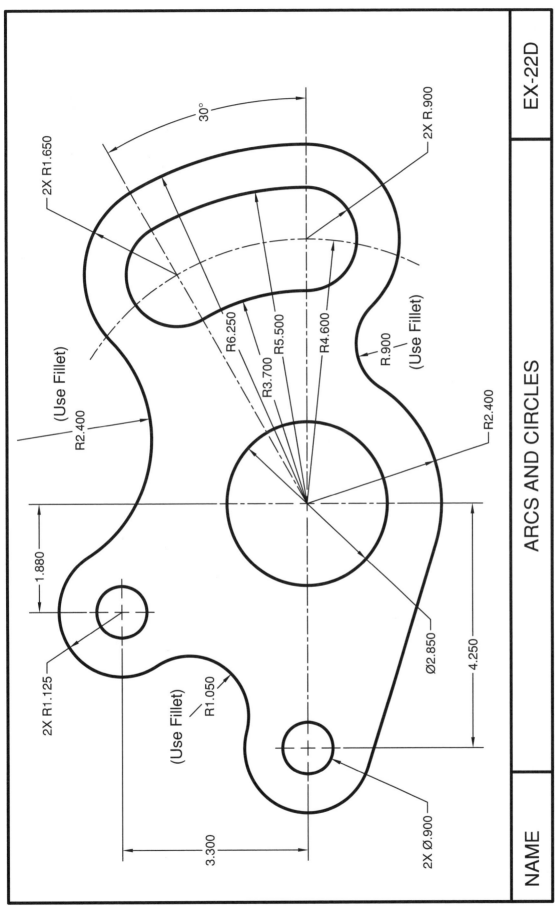

ARCS AND CIRCLES

EXERCISE 22D

EX-22D

NAME

INSTRUCTIONS:

1. Draw the Object above using ARCS, FILLET, CIRCLE and LINES.
2. Refer to lesson 18 if you need help with Dim Style settings for the Radial dims.
3. Refer to the next page for construction suggestions.
4. Save as: **EX-22D.**

Note: Helpful hints on the next page.

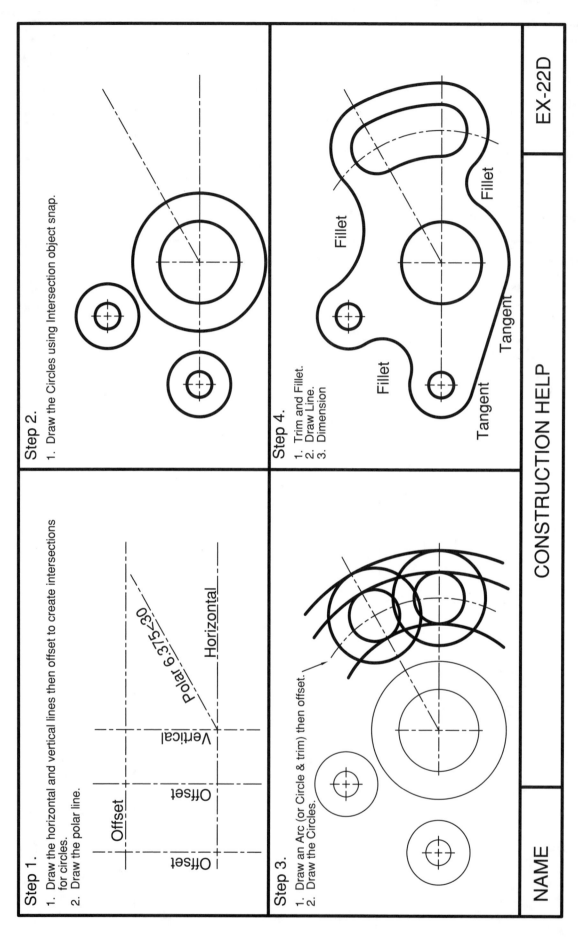

Step 1.
1. Draw the horizontal and vertical lines then offset to create intersections for circles.
2. Draw the polar line.

Offset

Offset

Vertical

Horizontal

Polar 6.375<30

Step 2.
1. Draw the Circles using Intersection object snap.

Step 3.
1. Draw an Arc (or Circle & trim) then offset.
2. Draw the Circles.

Step 4.
1. Trim and Fillet.
2. Draw Line.
3. Dimension

Fillet

Fillet

Fillet

Tangent

Tangent

Tangent

CONSTRUCTION HELP

EX-22D

NAME

EXERCISE 22D
helpful hints

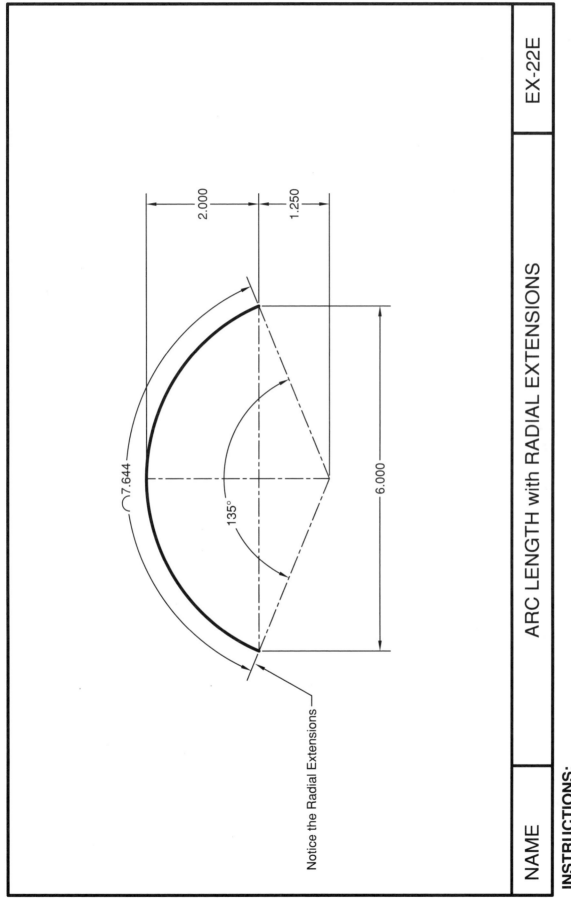

2.000

1.250

⌒7.644

135°

6.000

Notice the Radial Extensions

NAME	ARC LENGTH with RADIAL EXTENSIONS	EX-22E

EXERCISE 22E

INSTRUCTIONS:

1. Draw the Lines shown above.
2. Use the lines to draw the Arc. (You decide which Arc to use.)
3. Dimension as shown.
4. Save as: **EX-22E.**

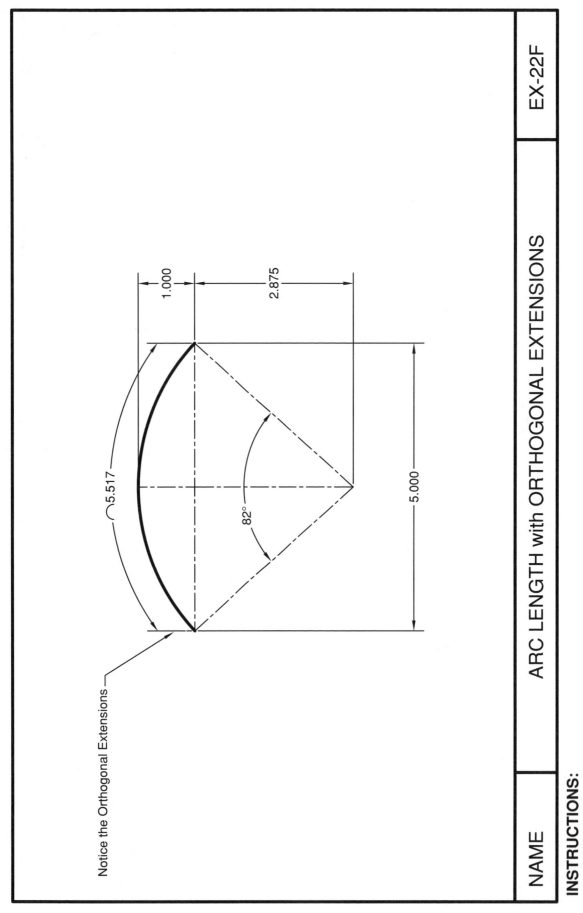

EXERCISE 22F

| NAME | ARC LENGTH with ORTHOGONAL EXTENSIONS | EX-22F |

1.000

2.875

5.000

⌒5.517

82°

5.000

Notice the Orthogonal Extensions

INSTRUCTIONS:

1. Draw the Lines shown above.
2. Use the lines to draw the Arc. (You decide which Arc to use.)
3. Dimension as shown.
4. Save as: **EX-22F.**

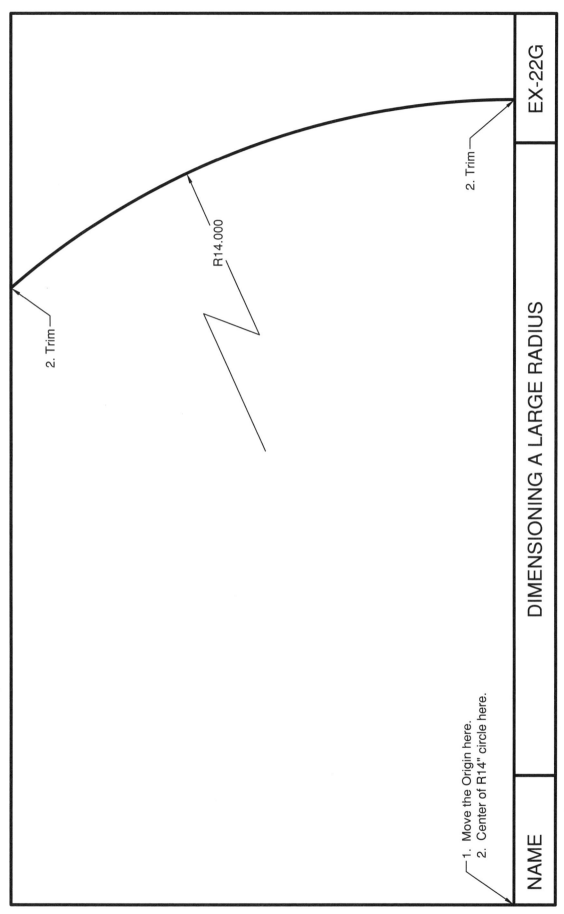

R14.000

2. Trim

2. Trim

1. Move the Origin here.
2. Center of R14" circle here.

NAME	DIMENSIONING A LARGE RADIUS	EX-22G

EXERCISE 22G

INSTRUCTIONS:

1. Move the Origin as shown.
2. Draw a Circle.   (Center = 0,0   Radius = 14") and trim it at the border lines as shown.
3. Create a "Jogged" dimension (R14.000) approximately as shown.
4. Save as: **EX-22G.**

NOTES:

LEARNING OBJECTIVES

After completing this lesson, you will be able to:

1. Understand what is a Polyline.
2. Draw a Polyline and Polyarc.
3. Assign widths to polylines.
4. Set the Fill mode to On or Off.

LESSON 23

POLYLINES

A **POLYLINE** is very similar to a LINE. It is created in the same way a line is drawn. It requires first and second endpoints. But a POLYLINE has additional features, as follows:

1. A **POLYLINE** is ONE object, even though it may have many segments.
2. You may specify a specific width to each segment.
3. You may specify a different width to the start and end of a polyline segment.

THE FOLLOWING ARE EXAMPLES OF POLYLINES WITH WIDTHS ASSIGNED.

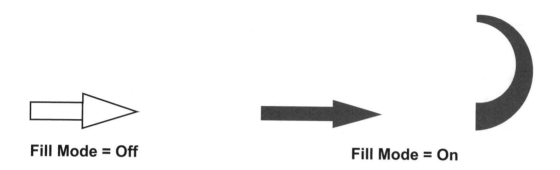

Fill Mode = Off   **Fill Mode = On**

To turn FILL MODE on or off.

1. Command: *type* FILL *<enter>*
2. Enter mode [On / Off] <ON>: *type ON or Off <enter>*
3. Command: *type REGEN <enter> or select: View / Regen*

NOTE: If you explode a POLYLINE it loses its width and turns into a regular line.

THE FOLLOWING IS AN EXAMPLE OF DRAWING A POLYLINE WITH "WIDTH"

1. Select the POLYLINE command using one of the following:

       TYPE = PL
       PULLDOWN = DRAW / POLYLINE
       TOOLBAR = DRAW

    Command: _pline
2. Specify start point: *place the first endpoint of the line*
    Current line-width is 0.000
3. Specify next point or [Arc/Halfwidth/Length/Undo/Width]: *type w* <enter>
4. Specify starting width <0.000>: *type the desired width* <enter>
5. Specify ending width <0.000>: *type the desired width* <enter>
6. Specify next point or [Arc/Close/Halfwidth/Length/Undo/Width]: *place the next endpoint*
7. Specify next point or [Arc/Close/Halfwidth/Length/Undo/Width]: *place the next endpoint*
8. Specify next point or [Arc/Close/Halfwidth/Length/Undo/Width]: *place the next endpoint*
9. Specify next point or [Arc/Close/Halfwidth/Length/Undo/Width]: *type C <enter>*

OPTIONS:

WIDTH
Specify the starting and ending width.

You can create a tapered polyline by specifying different starting and ending widths.

HALFWIDTH
The same as Width except the starting and ending halfwidth specifies half the width rather than the entire width.

ARC
This option allows you to create a circular polyline less than 360 degrees.

CLOSE
The close option is the same as in the Line command. Close attaches the last segment to the first segment.

LENGTH
This option allows you to draw a polyline at the same angle as the last polyline drawn. This option is very similar to the OFFSET command. You specify the first endpoint and the length. The new polyline will automatically be drawn at the same angle as the previous polyline.

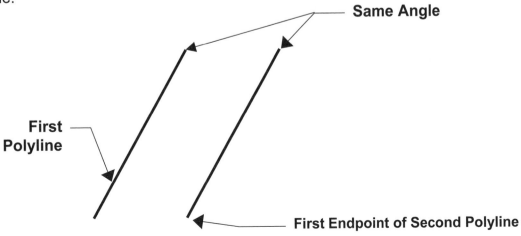

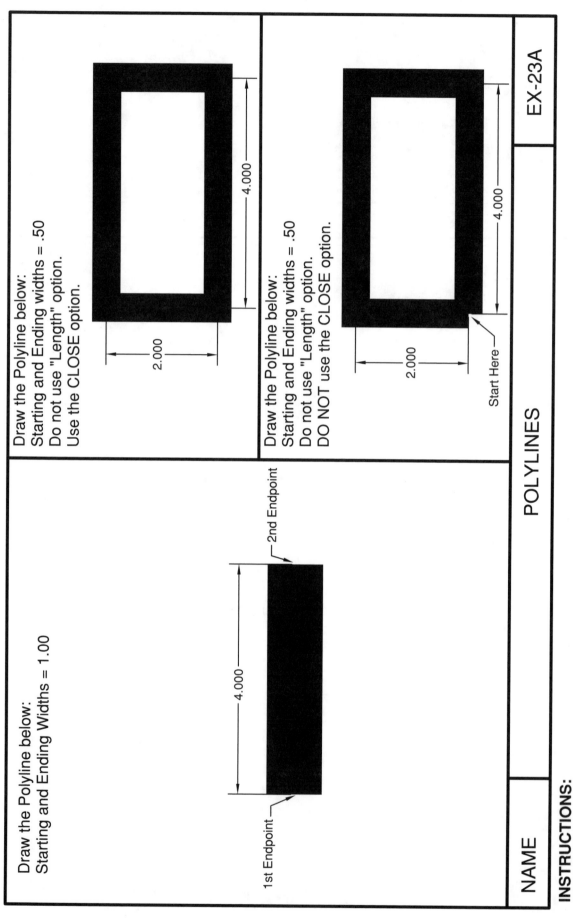

EXERCISE 23A

Draw the Polyline below:
Starting and Ending Widths = 1.00

1st Endpoint

2nd Endpoint

4.000

Draw the Polyline below:
Starting and Ending widths = .50
Do not use "Length" option.
Use the CLOSE option.

2.000

4.000

Draw the Polyline below:
Starting and Ending widths = .50
Do not use "Length" option.
DO NOT use the CLOSE option.

2.000

4.000

Start Here

EX-23A

POLYLINES

NAME

INSTRUCTIONS:

1. Draw the POLYLINES above. Follow the directions in each section.
   Do not dimension.

2. Save as: **EX-23A.**

INSTRUCTIONS:

Draw the object below as one continuous polyline as follows:

1. Select the polyline command.
2. Place the first endpoint.
3. Select "W" <enter>
      a. Starting width = 1.00
      b. Ending width = 1.00
4. Draw the segment 2" long.
5. Select "W" <enter>
      a. Starting width = 2.00
      b. Ending width = .50
6. Draw the segment 2" long.
7. Select "W" <enter>
      a. Starting width = .50
      b. Ending width = .50
8. Draw the last segment 2" long.

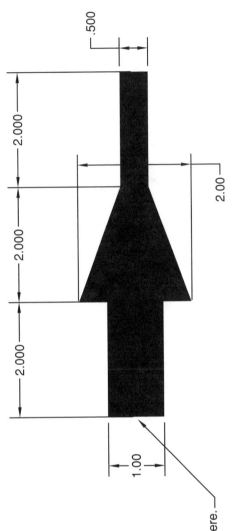

.500

2.000

2.000

2.000

2.00

1.00

Place 1st endpoint here.

NAME	CHANGING STARTING and ENDING WIDTHS	EX-23B

INSTRUCTIONS:

1. Follow the instructions shown above.
      Do not dimension.
2. Save as: **EX-23B.**

EXERCISE 23B

EXERCISE 23C

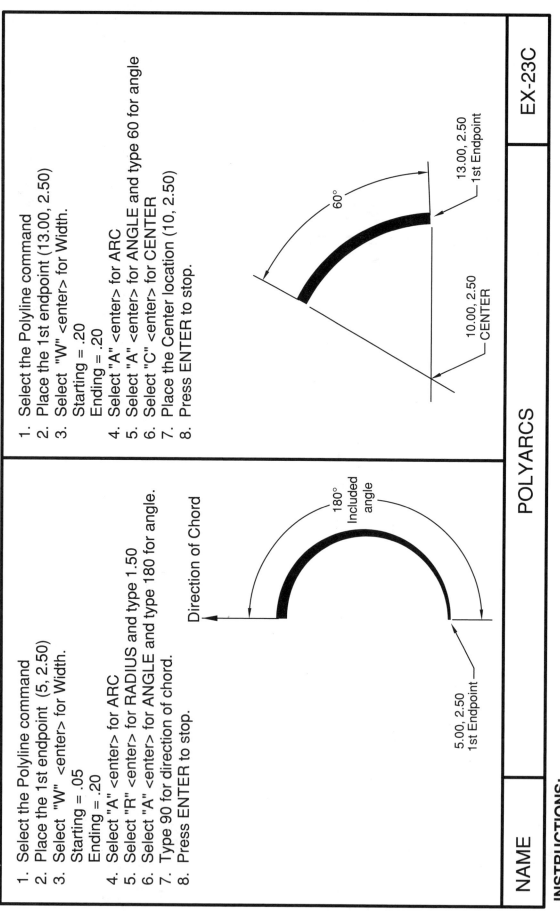

EX-23C

1. Select the Polyline command
2. Place the 1st endpoint (13.00, 2.50)
3. Select "W" <enter> for Width.
   Starting = .20
   Ending = .20
4. Select "A" <enter> for ARC
5. Select "A" <enter> for ANGLE and type 60 for angle
6. Select "C" <enter> for CENTER
7. Place the Center location (10, 2.50)
8. Press ENTER to stop.

60°

13.00, 2.50
1st Endpoint

10.00, 2.50
CENTER

POLYARCS

1. Select the Polyline command
2. Place the 1st endpoint (5, 2.50)
3. Select "W" <enter> for Width.
   Starting = .05
   Ending = .20
4. Select "A" <enter> for ARC
5. Select "R" <enter> for RADIUS and type 1.50
6. Select "A" <enter> for ANGLE and type 180 for angle.
7. Type 90 for direction of chord.
8. Press ENTER to stop.

180°
Included
angle

Direction of Chord

5.00, 2.50
1st Endpoint

NAME

INSTRUCTIONS:

1. Draw the POLYLINES above. Follow the direction in each section.
   Do not dimension.
2. Save as: **EX-23C.**

23-6

LEARNING OBJECTIVES

After completing this lesson, you will be able to:

1. Edit the width of a Polyline.
2. Join Polylines.
3. Convert a Polyline to a spline.
4. Convert a basic Line into a Polyline.
5. Join Lines, Arcs and Splines.

LESSON 24

EDITING POLYLINES

The **POLYEDIT** command allows you to make changes to a polyline's option, such as the width. You can also change a regular line into a polyline and JOIN the segments.

1. Select the **POLYEDIT** command using one of the following:

> **TYPE = PE**
> **PULLDOWN = MODIFY / OBJECT/ POLYLINE**
> **TOOLBAR =MODIFY II**

Note: You may modify "Multiple" polylines simultaneously.

2. PEDIT Select polyline or [Multiple]: *select the polyline to be edited or "M"*
3. Enter an option [Close/Join/Width/Edit vertex/Fit/Spline/Decurve/Ltypegen/Undo]: *select an Option (descriptions of each are listed below.)*

Note: If you select a line that is **NOT a POLYLINE**, the prompt will ask if you would like to turn it into a POLYLINE.

OPTIONS: (Step by step instructions in the following exercises)

CLOSE (Refer to Exercise 24A)
CLOSE connects the last segment with the first segment of an Open polyline. AutoCAD considers a polyline open unless you use the "Close" option to connect the segments originally.

OPEN (Refer to Exercise 24A)
OPEN removes the closing segment, but only if the CLOSE option was used to close the polyline originally.

JOIN (Refer to Exercise 24D)
The JOIN option allows you to join individual polyline segments into one polyline. The segments must have matching endpoints.

WIDTH (Refer to Exercise 24B)
The WIDTH option allows you to change the width of the polyline. But the entire polyline will have the same width.

EDIT VERTEX (Refer to Exercise 24C)
This option allows you to change the starting and ending width of each segment individually.

SPLINE (Refer to Exercise 24D)
This option allows you to change straight polylines to curves.

DECURVE
This option removes the SPLINE curves and returns the polyline to its original straight line segments.

JOIN COMMAND

On the previous page the Join option within the polyedit command is explained. That option can only be used to join individual polyline segments.

The JOIN command can join Lines, Polylines, Arcs and Splines.

1. Select the **JOIN** command using one of the following:

    TYPE = J
    PULLDOWN = MODIFY / JOIN
    TOOLBAR =MODIFY ⊣⊢

2. Select the Source object.

3. Select the objects to join to the source object.

Seems easy, and it is – but there are a few rules regarding each type of object you must understand.

LINES
The Lines must be collinear (on the same infinite line) but can have gaps between them.

POLYLINES
Same as Lines except no gaps allowed between them.

ARC
The Arcs must line on the same imaginary circle but can have gaps.
The close option will turn them into a circle.

SPLINE
The Spline must lie in the same plane, and contiguous (end to end).
(You haven't learned this command as yet)

EXERCISE 24A

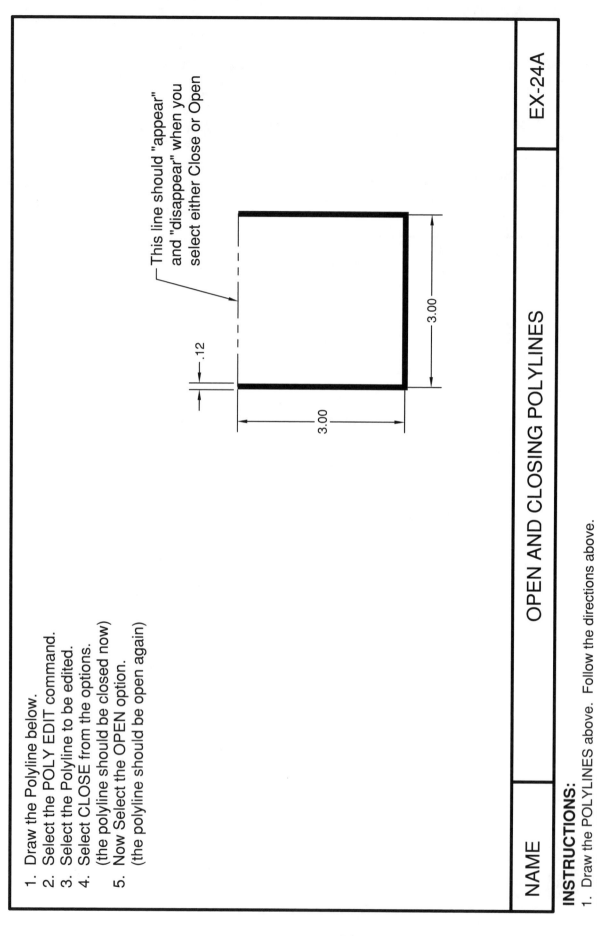

1. Draw the Polyline below.
2. Select the POLY EDIT command.
3. Select the Polyline to be edited.
4. Select CLOSE from the options.
   (the polyline should be closed now)
5. Now Select the OPEN option.
   (the polyline should be open again)

This line should "appear" and "disappear" when you select either Close or Open

.12

3.00

3.00

OPEN AND CLOSING POLYLINES

NAME

INSTRUCTIONS:

1. Draw the POLYLINES above. Follow the directions above.
2. Save as: **EX-24A.**

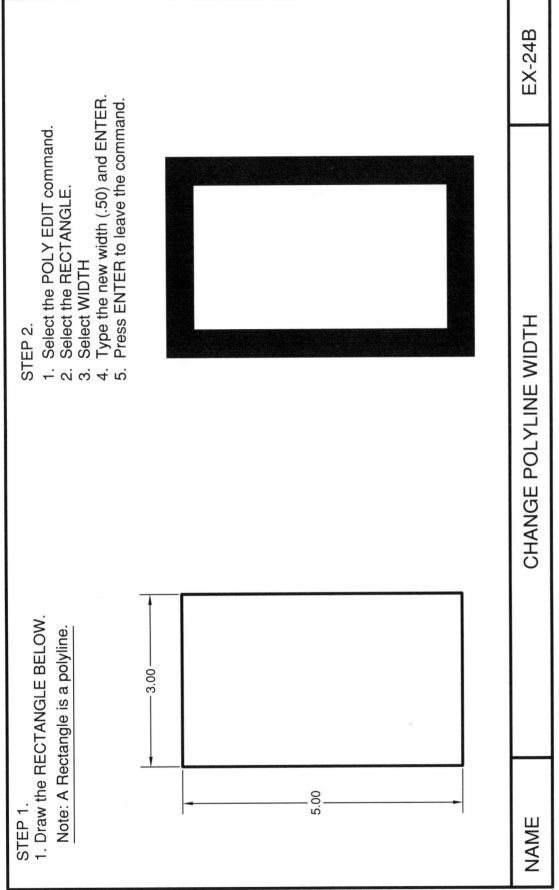

STEP 1.

1. Draw the RECTANGLE BELOW.

Note: A Rectangle is a polyline.

STEP 2.

1. Select the POLY EDIT command.
2. Select the RECTANGLE.
3. Select WIDTH
4. Type the new width (.50) and ENTER.
5. Press ENTER to leave the command.

3.00

5.00

EXERCISE 24B

CHANGE POLYLINE WIDTH

NAME

INSTRUCTIONS:

1. Draw the RECTANGLE above. Follow the directions above.
2. Save as: **EX-24B.**

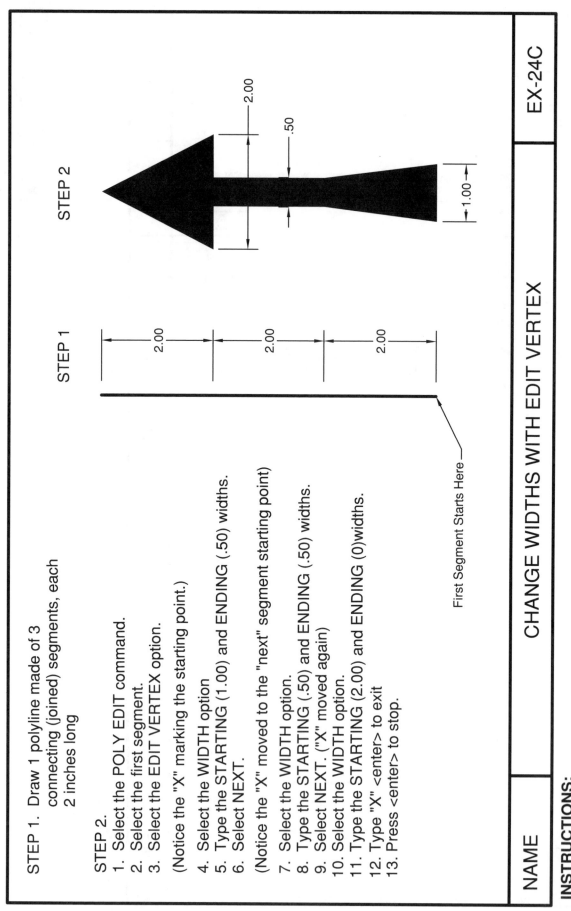

STEP 2

STEP 1

2.00

2.00

.50

2.00

1.00

2.00

STEP 1. Draw 1 polyline made of 3
connecting (joined) segments, each
2 inches long

STEP 2.
1. Select the POLY EDIT command.
2. Select the first segment.
3. Select the EDIT VERTEX option.

(Notice the "X" marking the starting point.)

4. Select the WIDTH option
5. Type the STARTING (1.00) and ENDING (.50) widths.
6. Select NEXT.

(Notice the "X" moved to the "next" segment starting point)

7. Select the WIDTH option.
8. Type the STARTING (.50) and ENDING (.50) widths.
9. Select NEXT. ("X" moved again)
10. Select the WIDTH option.
11. Type the STARTING (2.00) and ENDING (0)widths.
12. Type "X" <enter> to exit
13. Press <enter> to stop.

First Segment Starts Here

EXERCISE 24C

EX-24C

CHANGE WIDTHS WITH EDIT VERTEX

NAME

INSTRUCTIONS:

1. Draw the POLYLINE above. Follow the directions above.
2. Save as: **EX-24C.**

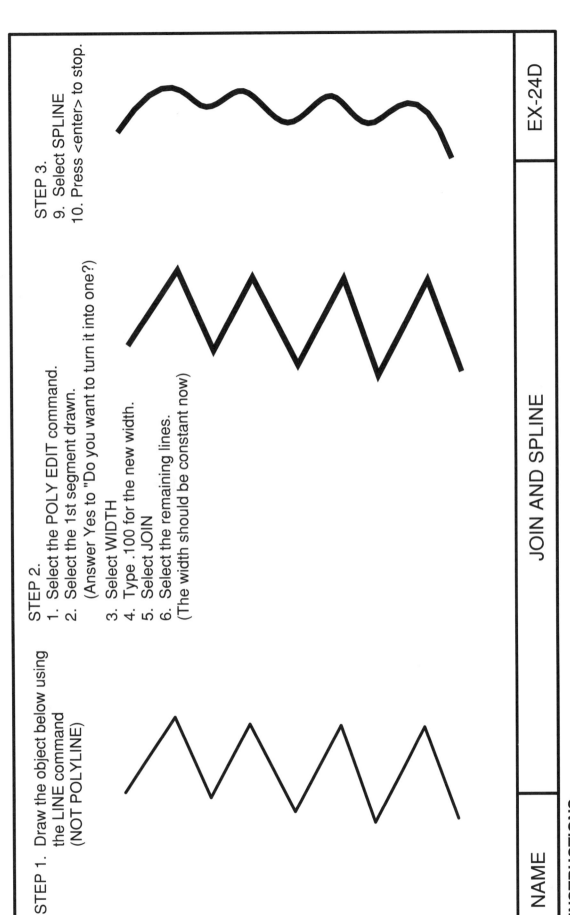

STEP 1. Draw the object below using the LINE command. (NOT POLYLINE)

STEP 2.
1. Select the POLY EDIT command.
2. Select the 1st segment drawn.
   (Answer Yes to "Do you want to turn it into one?")
3. Select WIDTH
4. Type .100 for the new width.
5. Select JOIN
6. Select the remaining lines.
   (The width should be constant now)

STEP 3.
9. Select SPLINE
10. Press <enter> to stop.

JOIN AND SPLINE

EX-24D

NAME

EXERCISE 24D

INSTRUCTIONS:
1. Draw the POLYLINES above. Follow the directions above.
2. Save as: **EX-24D.**

EXERCISE 24E

INSTRUCTIONS:

1. Draw 3 Lines as shown below.
   The Lines must all be on the same "Y" axis.
2. Select the JOIN command.
3. Select the left line. (Source Line)
4. Select the remaining 2 Lines.
5. Press <enter>

The 3 lines should now be one continuous Line.

SOURCE LINE

JOIN MULTIPLE UNATTACHED LINES

EX-24E

NAME

INSTRUCTIONS:

1. Draw the POLYLINES above. Follow the directions above.
2. Save as: **EX-24E.**

24-8

LEARNING OBJECTIVES

After completing this lesson, you will be able to:

1. Create a new Text Style.
2. Change an Existing Text Style.
3. Change the Point Style.
4. Place a Point at designated intervals on an object.

LESSON 25

CREATING NEW TEXT STYLES

AutoCAD provides you with only one Text Style named "Standard". You may want to create a new text style with a different font and effects. Steps 1 through 8 below will guide you through the process.

1. Select the TEXT STYLE command using one of the following:
   TYPE = STYLE or ST
   PULLDOWN = FORMAT / TEXT STYLE
   TOOLBAR = FORMAT

   The TEXT STYLE dialog box below should appear.

The information in this dialog box is a description of the Text Style highlighted in the Style Name box.

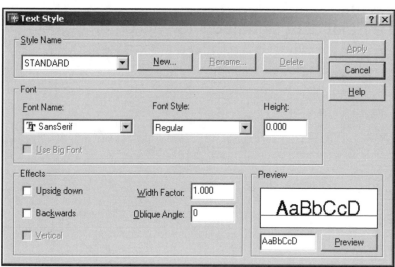

2. Select the **NEW** button.

4. Select the **FONT**.

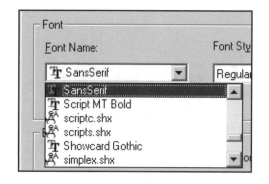

3. Type the new style name in **STYLE NAME** box. Then select the **OK** button.

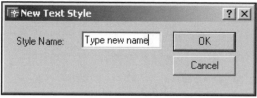

Text Styles can have a maximum of 31 characters, including letters, numbers, dashes, underlines and dollar signs. You can use Upper or Lower case.

5. Enter the value of the Height. **(If the value is 0, AutoCAD will always prompt you for a height. If you enter a number the new text style will have a fixed height and AutoCAD will not prompt you for the height)**

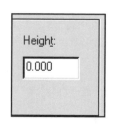

6. Assign **EFFECTS.**

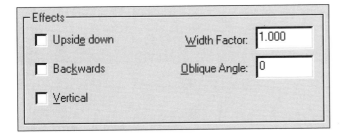

UPSIDE-DOWN
Each letter will be created upside-down in the order in which it was typed.
(Note: this is different from rotating text 180 degrees.)

BACKWARDS
The letters will be created backwards as typed.

VERTICAL
Each letter will be inserted directly under the other. Only **.shx** fonts can be used.
VERTICAL text will not display in the **PREVIEW** box.

OBLIQUE ANGLE
Creates letter with a slant, like italic. An angle of 0 creates a vertical letter. A positive angle will slant the letter forward. A negative angle will slant the letter backward.

WIDTH FACTOR
This effect compresses or extends the width of each character. A value less than 1 compresses. A value greater than 1 extends each character.

7. **PREVIEW** your settings.

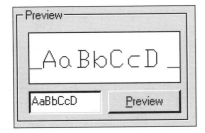

8. Select the **APPLY** button and then the **CLOSE** button.

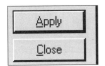

CREATING YOUR NEW TEXT STYLE IS NOW COMPLETE.

CHANGING TEXT STYLES

RENAMING

1. Select the **TEXT STYLE** command.
   The Text Style Dialog box appears.

2. Select the style
   you want to rename.

3. Click on
   RENAME button.

4. Type the New name then
   click on the **OK** button.

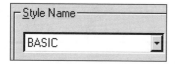

5. Click on the **CLOSE** button.

DELETING

1. Select the **TEXT STYLE** command.
   The Text Style Dialog box appears.

2. Select the style
   you want to DELETE.

3. Click on the
   DELETE button.

4. Warning appears, select
   Yes or No.

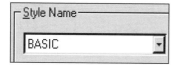

5. Click on the **CLOSE** button.

CHANGING EFFECTS

1. Select the **TEXT STYLE** command.
   The Text Style Dialog box appears.

2. Make the changes.

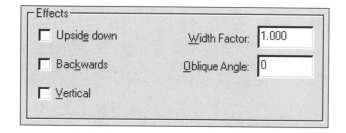

3. Click on the **APPLY** button and then **CLOSE**.

DIVIDE

The DIVIDE command divides an object mathematically by the NUMBER of segments you designate. It then places a POINT (object) at each interval on the object.

Note: the object selected is **NOT** broken into segments. The POINTS are simply drawn **ON** the object.

First select the **POINT STYLE** to be placed on the object. *Refer to lesson 5*

Next select the **DIVIDE** command using one of the following:

> **TYPE = DIV**
> **PULL DOWN = DRAW / POINT / DIVIDE**
> **TOOLBAR = DRAW**

Select object to divide: *select the object to divide.*
Enter the number of segments or [Block]: *type the number of segments <enter>*

EXAMPLE:

This LINE has been DIVIDED into 4 EQUAL lengths.
But remember, the line is not broken into segments.
The Points are simply drawn ON the object.

MEASURE

The **MEASURE** command is very similar to the **DIVIDE** command because point objects are drawn at intervals on an object. However, the **MEASURE** command allows you to designate the **LENGTH** of the segments rather than the number of segments.

Note: the object selected is **NOT** broken into segments. The **POINTS** are simply drawn **ON** the object.

First select the **POINT STYLE** to be placed on the object. *Refer to lesson 5*

Next select the **MEASURE** command using one of the following:

> **TYPE = ME**
> **PULL DOWN = DRAW / POINT / MEASURE**
> **TOOLBAR = DRAW**

Command: _measure
Select object to measure: *select the object to measure.*
(Note: this selection point is also where the MEASUREment will start.)
Specify length of segment or [Block]: *type the length of one segment <enter>*

EXAMPLE:

Select Object here to start the measurement from the left.

The MEASUREment was started at the left endpoint, and ended just short of the right end of the line. The remainder is less than the measurement length designated.

EXERCISE 25A

INSTRUCTIONS:

1. Open your "**BSIZE**" drawing.
2. Create the 3 text styles listed below.
3. Save as EX-25A (The text is not visible, but it will be after completing EX-25B)

Follow the instruction **"CREATING NEW TEXT STYLES"** on page 25-2.

Name:	Style1
Font:	Romand.shx
Height:	.50
Upside down:	No
Backwards:	No
Vertical:	Yes
Width factor:	1
Oblique angle:	30

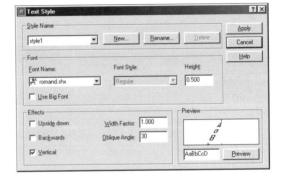

Name:	Style2
Font:	gothice.shx
Height:	1.00
Upside down:	No
Backwards:	Yes
Vertical:	No
Width factor:	.75
Oblique angle:	0

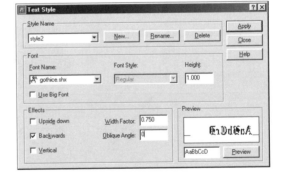

Name:	Style3
Font:	Italict.shx
Height:	2.00
Upside down:	Yes
Backwards:	No
Vertical:	No
Width factor:	.75
Oblique angle:	0

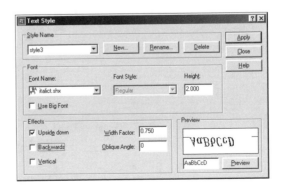

START POINT 5, 8

START POINT 4, 5

START POINT 7, 1.5

| NAME | DRAWING WITH NEW TEXT STYLES | EX-25B |

EXERCISE 25B

INSTRUCTIONS:

1. Open drawing 25A.
2. Draw the text styles above using the text style you created.
3. Select the appropriate text style before typing.
4. Notice "Start Point" locations.
5. Save as: **EX-25B and plot.**

STYLE1

STYLE3

STYLE2

PROPERTIES

Text

General

Color	White
Layer	OBJECT
Linetype	ByLayer
Linetype...	1.000
Plot style	ByColor
Lineweight	ByLayer
Hyperlink	
Thickness	.000

Text

Contents	STYLE3
Style	style3
Justify	STANDARD
Height	style1
Rotation	style2
Width fa...	style3
	SSBOLD
Obliquing	0
Text alig...	1.475
Text alig...	-.250
Text alig...	.000

Specifies the style name of the text

EX-25C

NAME | CHANGING THE TEXT STYLE

EXERCISE 25C

INSTRUCTIONS:

1. Open drawing 25B.
2. Using MODIFY / PROPERTIES change the 3 text styles to Text Style "Standard".
3. Note: you must change one at a time and "deselect" between style changes.
   If you do not deselect between style changes, you will not acheive the same results.
4. Save as: **EX-25C and plot.**

EXERCISE 25D

INSTRUCTIONS:

1. Open File **25C.**
2. Delete text styles **1,2 and 3.**

There are 2 methods. Try both.

METHOD 1.

1. Select **FORMAT / TEXT STYLE**
2. Select the "Standard" text style.
3. Select the CLOSE button. (This makes Standard the current text style)
4. Now return to **FORMAT / TEXT STYLE**
5. Select the style you want to **delete**.
6. Click on the **DELETE** button.

AutoCAD will not allow you to delete an active or current text style. That is why you must make the "Standard" text style current before you may delete a text style. (steps 2 and 3 shown above)

METHOD 2.

1. Select **File / Drawing Utilities / Purge**

2. Select the **+ sign** beside Text Styles
   (It will turn to a **– sign** as shown)

3. Select the Style 1, 2 and 3

***(Hold the shift key down to select
 more than one item)***

4. Select **Purge** button.

5. Select **Close**.

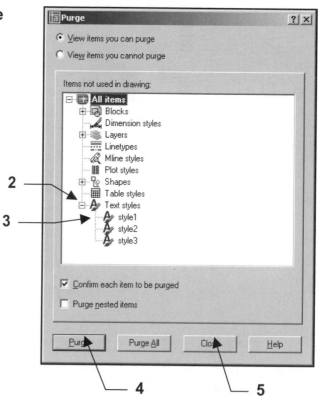

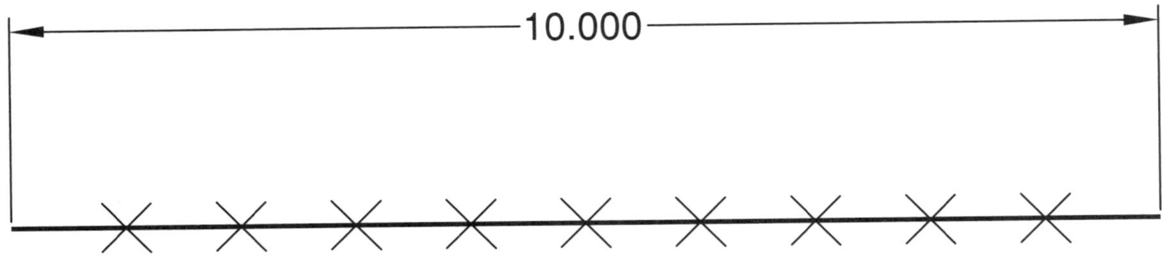

EXERCISE 25E

NAME	DIVIDE	EX-25E

10.000

INSTRUCTIONS:

1. Draw a LINE 10 inches long.
2. Set the POINT STYLE to **X** using: FORMAT / POINT STYLE.
3. DIVIDE the line into 10 equal divisions using: **a**. DRAW / POINT / DIVIDE.
   b. *Select the object to divide.*   **c**. *Type 10 for number of segments.*
4. Save as: **EX-25E and plot.**

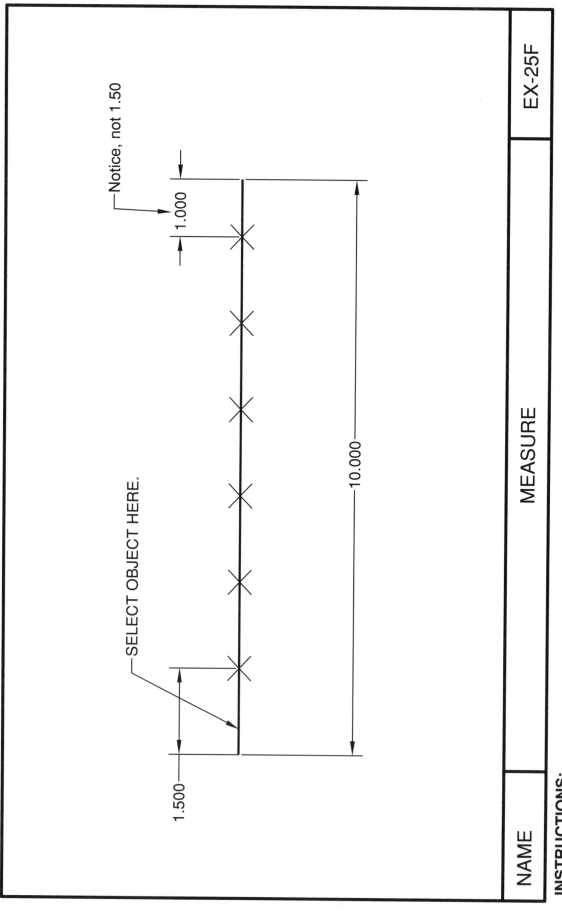

EXERCISE 25F

NAME	MEASURE	EX-25F

INSTRUCTIONS:

1. Draw a LINE 10 inches long.
2. Set the POINT STYLE to **X** using: FORMAT / POINT STYLE.
3. Place POINTS at a measurement of 1.50 using: **a**. DRAW / POINT / MEASURE.
   b. *Select the location to start the measurement from.*   **c**. *Type 1.50 for the measurement.*
4. Save as: **EX-25F and plot.**

EXERCISE 25G

3X Ø1.00 EQUALLY SPACED

4.000

6.750

1.500
1.500
1.500
1.500
1.500
1.500

4X Ø.75

8X Ø.50
EQUALLY SPACED

MORE DIVIDE and MEASURE

NAME

INSTRUCTIONS:

1. Draw the RECTANGLE and the DIAGONAL LINE.
2. EXPLODE the Rectangle and set the POINT STYLE.
3. Place POINTS as shown using: DIVIDE or MEASURE.
4. Draw the Circles using NODE snap to locate the center locations.
5. Dimension and Save as: **EX-25G and plot.**

NOTES:

LEARNING OBJECTIVES

After completing this lesson, you will be able to:

1. Create your own Layers.
2. Load Linetypes into your drawing.
3. Understand the difference between Model and Layout tabs.
4. Adjust the size of the "Pick" box.
5. Create Floating Viewports
6. Create a Page Setup for plotting your drawings.
7. Create a Decimal "Setup" master file for future use.
8. Create a new Border to use when plotting.

LESSON 26

SERIOUS BUSINESS in this lesson

In the previous lessons you have been having fun learning most of the basic commands that AutoCAD offers and you have been using our drawings that include preset layers, text styles, drawing settings etc. But now it is time to get down to the "serious business" of <u>setting up your own drawing from "scratch"</u>.

<u>Starting from scratch</u> means you will need to set or create the following:

<u>Items 1 through 4 you have learned in previous lessons</u>
1. Drawing Units (Lesson 4)
2. Snap and Grid (Lesson 2)
3. Create Text styles (Lesson 25)
4. Create Dimension Styles (Lesson 16)

<u>Items 5 through 9 will be learned in this lesson</u>
5. Create new layers and load linetypes.
6. Create a "Layout" for plotting.
7. Create a "Floating Viewport" in the Layout.
8. Create a "Page Setup" to save plot settings.
9. Plot the drawing from Paper space.

*After reading pages 26-3 through 26-18 start Exercise 26A and work your way through to 26D. When you have completed Exercise 26D you will have created a master drawing named, "**<u>My Decimal Setup</u>**". This master drawing will have everything set, created and prepared, ready to use each time you want to create a drawing using decimal units and to be plotted on an 11 x 17 inch sheet.*

This means, for future drawings you merely open "My Decimal Setup" and start drawing. No time consuming setups. It is all ready to go.

In Lesson 27 you will create a master drawing for "feet and inches".

So take it one page at a time and really concentrate on understanding the process.

CREATING NEW LAYERS

Using layers is an important part of managing and controlling your drawing. It is better to have too many layers than too few.   Draw like objects on the same layer.
For example, place all doors on the layer "door" or centerlines on the layer "centerline".
When you create a new layer you will assign a name, color, linetype, lineweight and whether or not it should plot.

1. Select the Layer command using one of the following:

> **TYPE = LA**
> **PULLDOWN = FORMAT / LAYER**
> **TOOLBAR = OBJECT PROPERTIES**

The Layer Properties Manager dialog box shown below will appear.

2. NEW LAYER button

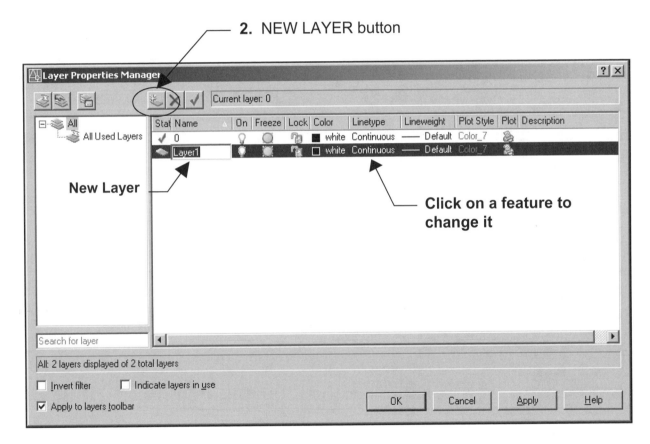

New Layer

Click on a feature to change it

2. Select the New Layer button and a new layer will appear.
    Type the new layer name and press <enter>

3. Click on any of the features and a dialog box will appear.

**The following pages will describe <u>Color</u>, <u>Linetype</u>, and <u>Layer Controls</u>.
Refer to page 9-7 for a reminder on <u>Lineweights</u>.**

COLOR

Color is not merely to make a pretty display on the monitor screen.

Here are some things to consider when selecting the colors for your layers.

a. Now that color printers are so commonplace, consider how the colors will appear on the paper. (Pastels do not appear well on white paper.)

b. Consider how the colors will appear on the screen. (Yellow appears well on a black background but not on white. Changing the screen color is in the Advanced Workbook.)

c. If you choose to use color dependent plot styles instead of lineweights, color will determine the width of the lines.

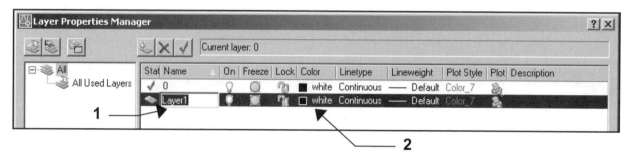

1. Select (highlight) the layer that you want to change.

2. Select the color swatch or word. (white)

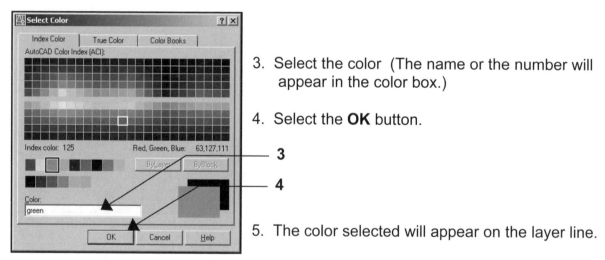

3. Select the color (The name or the number will appear in the color box.)

4. Select the **OK** button.

5. The color selected will appear on the layer line.

LOADING and SELECTING a LINETYPE

In an effort to conserve data within a drawing file, AutoCAD automatically loads only one linetype called "continuous". If you would like to use other linetypes, such as "dashed", you must "Load" them into the drawing as follows.

1. Select the Linetype.

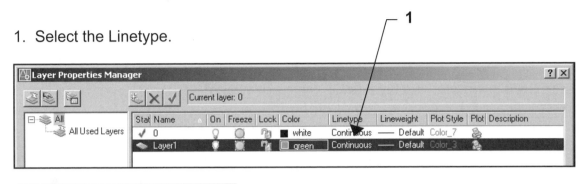

2. Select the **LOAD** button.

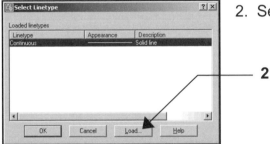

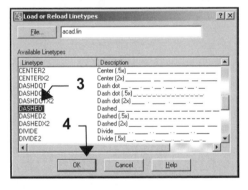

3. Scroll down the list and select a linetype. (Hold the **Ctrl** button down to select multiple linetypes)

4. Select the **OK** button.

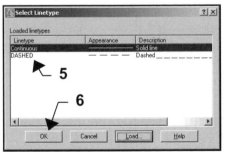

The linetype appears in the loaded linetypes list.

5. Select the linetype to assign to the layer.

6. Select the **OK** button.

7. The linetype appears on the layer line.

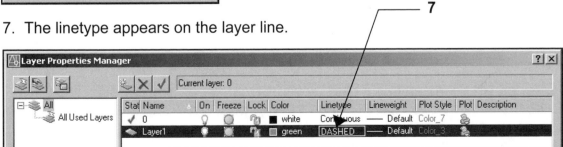

LAYER CONTROL DEFINITIONS

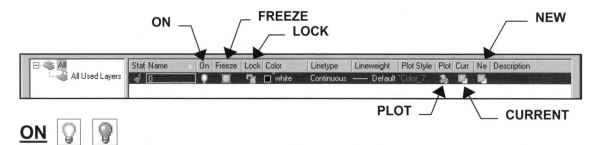

ON

If a layer is **ON** it is **visible**. If a layer is **OFF** it is **not visible**.
Only layers that are **ON** can be **edited** or **plotted**.
(Warning: Objects on a Layer that is OFF can be accidentally erased even though they are invisible. When you are asked to select objects in the erase command, if you type **ALL** <enter> all objects will be selected; even the invisible ones.)

LOCK

LOCKED layers are visible but cannot be edited. They are visible so they **will** be plotted. (Locked layers cannot be selected by typing ALL.)

PLOT

This option prevents a layer from plotting even though it is visible.

FREEZE

Freeze and **Thaw** are very similar to On and Off. A Frozen layer is not visible.
A Thawed layer is visible. Only thawed layers can be edited or plotted.

Additionally:
a. Objects on a Frozen layer **cannot** be accidentally erased by typing All.
b. When working with large and complex drawings, freezing saves time because frozen layers are not **regenerated** when you zoom in and out.

Note: The next 2 are displayed only when you select a Layout tab.

CURRENT

Layers selected will be frozen in the "current" viewport only. Current means the active viewport. Only one viewport can be active at one time.
(Available in Paper Space only)

NEW

Layers selected will be frozen in the next viewport created. So you are selecting the layers to be frozen before you have created the viewport. This one doesn't get used much.
(Available in Paper Space only)

MODEL and LAYOUT tabs

Read this information carefully. It is very important that you understand this concept.

AutoCAD provides two drawing spaces, **MODEL** and **LAYOUT**. You move into one or the other by selecting either the MODEL or LAYOUT tabs, located at the bottom left of the drawing area.

<u>Model Tab</u> (Also called *Model Space*)
When you select the Model tab you enter <u>MODEL SPACE</u>.
(This is where you have been drawing and plotting from for the last 25 lessons)
Model Space is where you **create** and **modify** your drawings.

<u>Layout1 Tab</u> (Also called *Paper Space*)

When you select a Layout tab you enter <u>PAPER SPACE</u>.
The primary function of Paper Space is to prepare the drawing for plotting.

When you select the Layout tab for the first time, the "<u>Page Setup Manager</u>" dialog box will appear. Using this dialog box, you will assign a name for your new page setup. Then you will specify which plotting device and paper size to use for plotting. For now, select the CLOSE button to continue.
(More information on this in "How to create a Page Setup" page 26-12)

Notice that Model Space seems to have disappeared, and a <u>blank sheet of paper</u> is displayed on the screen. This sheet of paper is basically in front of the Model Space. (See illustration below) To see your drawing (Model Space) while still in Paper Space, you must <u>cut a hole</u> in this sheet. This hole is called a **"Viewport"**. *(Refer to "Viewports" page 26-8.)*

<u>Try to think of this as a picture frame (paper space) in front of a photograph (model space).</u>

Generally, the only objects that should be in paper space are the Title Block, Border, Dimensions and Notes.

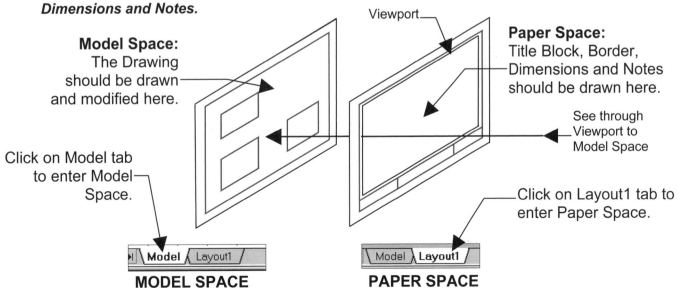

Model Space:
The Drawing
should be drawn
and modified here.

Click on Model tab
to enter Model
Space.

Viewport

Paper Space:
Title Block, Border,
Dimensions and Notes
should be drawn here.

See through
Viewport to
Model Space

Click on Layout1 tab to
enter Paper Space.

MODEL SPACE   **PAPER SPACE**

VIEWPORTS

Viewports are only used in Paper Space (Layout tab).
Viewports are holes cut into the sheet of paper displayed on the screen.
Viewports are objects. They can be moved, stretched, scaled, copied and erased.
You can have multiple Viewports, and their size and shape can vary.

Note: It is considered good drawing management to create a layer for the Viewport "frames" to reside on. This will allow you to control them separately; such as setting the viewport layer to "No plot" so it will not be visible when plotting.

HOW TO CREATE A VIEWPORT

1. First, create your drawing in Model Space (Model tab) and save it.

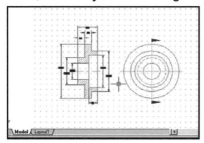

2. Select the "Layout1" tab. (If the "Page Setup Manager" dialog box appears, select the **CLOSE** button for now or refer to "How to Create a Page setup" on page 26-12.)
3. You are now in Paper Space. Model Space appears to have disappeared, but a blank paper is actually in front of Model Space, preventing you from seeing your drawing. (Your Border, title block and notes will be drawn on this paper in Paper Space.)

4. Select layer "Viewport" (You may need to create one, refer to 26-3)
5. Open the "Viewports" toolbar (Refer to page 2-9)

2006 toolbar

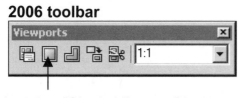

2006 LT toolbar

6. Select the "Single Viewport" icon.

7. Draw the Viewport "frame" by specifying the location for the "first corner" and then the "opposite corner". (Similar to drawing a Rectangle, but **do not** use the Rectangle command.)

You should now be able to look through the Paper Space sheet to Model Space and see your drawing.

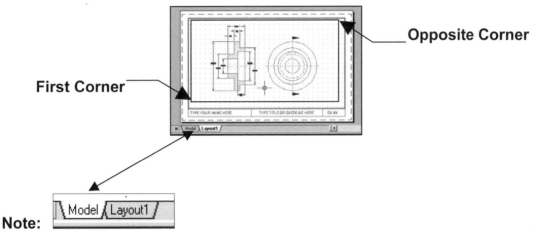

Opposite Corner

First Corner

Note:
You may go back to Model Space simply by clicking on the Model tab.
To return to Paper Space, click on the Layout tab.
(Make sure your grids are ON in Model Space and OFF in Paper Space. Otherwise you will have double grids)

EXAMPLE OF WHY WE WANT TO USE PAPER SPACE

I know you are probably wondering why you should bother with Paper Space. Paper Space is a great tool to manipulate your drawing for plotting.
Notice the drawing below. Viewport 1 displays the entire drawing. In Viewport 2 the scale has been adjusted to get a closer look at that section arrow. Viewport 3 not only has it's scale adjusted but the dimension layer is frozen (invisible) in the "current viewport" only. Notice the dimension layer is still thawed (visible) in the upper viewport. Can you guess which is the **Active** viewport at the moment? (That's right; Viewport 3)

Experiment and get familiar with Paper Space. It is a <u>very useful</u> tool.

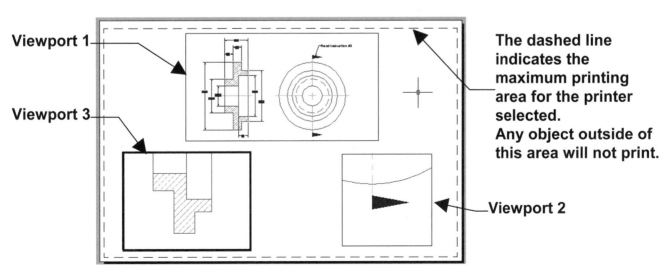

Viewport 1

Viewport 3

The dashed line indicates the maximum printing area for the printer selected.
Any object outside of this area will not print.

Viewport 2

TO REACH THROUGH TO MODEL SPACE WHILE IN PAPER SPACE

Note: You may draw in model space (inside the viewport) while in Paper Space (Layout tab). You may switch to Model or Paper Space using one of the following methods.

Method 1.
Select the **Model / Paper** button on the Status Line.

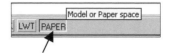

Method 2.
If you double click <u>inside</u> the viewport frame, it will activate the Model Space.
The Model space UCS icon will appear within the viewport and the frame will appear heavy.

Model Space UCS icon displayed

Double click insid the Viewport to activate Model Space.

Viewport "frame" appears heavy.

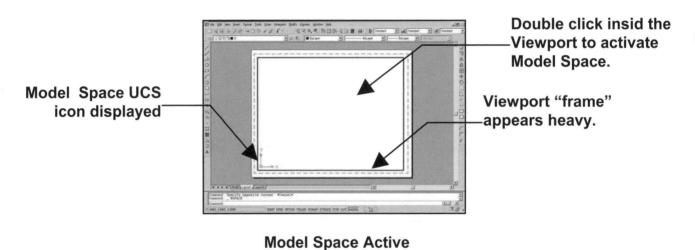

Model Space Active

To return to Paper Space, double click in the gray area and the Paper Space icon will appear.

Double click in the gray area to activate Paper Space.

Paper Space Icon

Viewport "frame" appears thin.

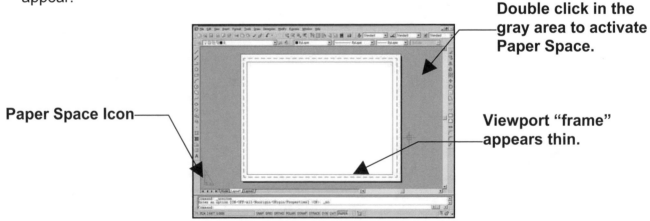

Paper Space Active

VIEWPORTS TOOLBAR TOOLS

TOOLBAR FOR <u>2006</u> SHOWN BELOW: (Refer to How to open toolbars 2-9)

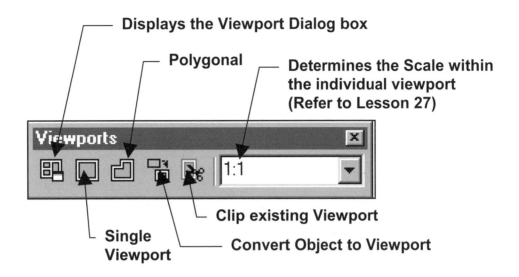

Displays the Viewport Dialog box

Polygonal

Determines the Scale within the individual viewport (Refer to Lesson 27)

Single Viewport

Clip existing Viewport

Convert Object to Viewport

TOOLBAR FOR <u>LT</u> SHOWN BELOW:

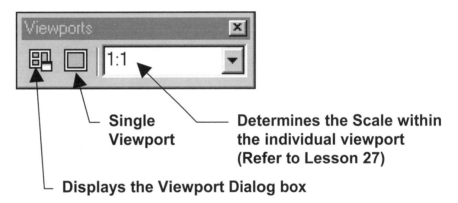

Single Viewport

Determines the Scale within the individual viewport (Refer to Lesson 27)

Displays the Viewport Dialog box

HOW TO LOCK A VIEWPORT

After you have adjusted the scale within each viewport, you will want to "LOCK" the viewport so it will not change. <u>Here are 3 methods</u>.

Method 1. a. Make sure you are in Paper Space.
      b. Left click <u>once</u> on the Viewport frame
      c. Right Click (the short cut menu should appear)
      d. Select "**Display Locked**" - Select Yes or No.

Method 2. a. Left click <u>once</u> on the Viewport frame
      b. Right Click and select "Properties".
      c. Select **Misc / Displayed Locked**. (Select Yes or No)

Method 3. a. Type **MV** at the command line and press <enter>
      b. Select "Lock" <enter>
      b. Type On or Off <enter>
      c. Click on the Viewport frame.

HOW TO CREATE A PAGE SET UP for PAPER SPACE

To plot a drawing from Paper Space, you must first specify what <u>printing device</u> to use and what <u>paper size</u> to print on. These specifications are called the "**Page Setup**".

Note: The following is for concept only. The actual exercise starts with 26A.

STEP 1. OPEN THE DRAWING YOU WISH TO PLOT.
   The drawing must be displayed on the screen.

STEP 2. SET UP THE PAPER SPACE ENVIRONMENT.
   Setting up the Paper Space environment means selecting the printer and specifying paper size.

A. Select a **LAYOUT** tab.

Note: If the "Page Setup Manager" dialog box shown below does not appear automatically, right click on the Layout tab, then select Page Setup Manager.

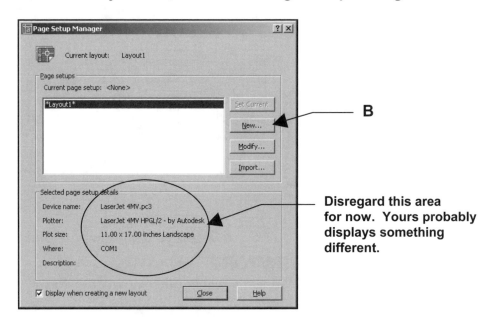

Disregard this area for now. Yours probably displays something different.

B. Select the **New** button. *(The "New Page Setup" dialog should appear)*

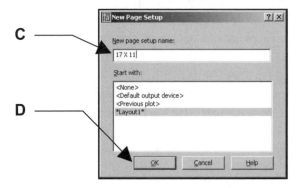

C. Type the new page setup name.
D. Select the **OK** button.

The "Page Setup" dialog box should appear.

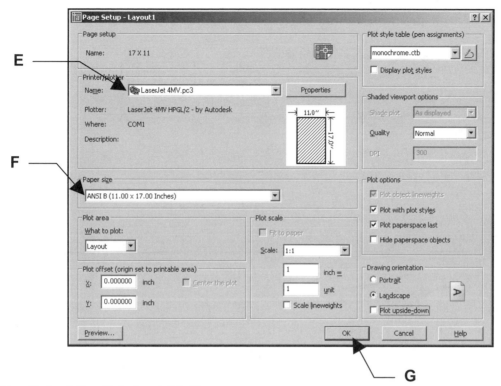

E. Select the **Printer / Plotter**.
F. Select the **Paper Size.**
G. Select the **OK** button.

The "Page Setup Manager" should have returned. Notice the new page setup name now appears in the list.

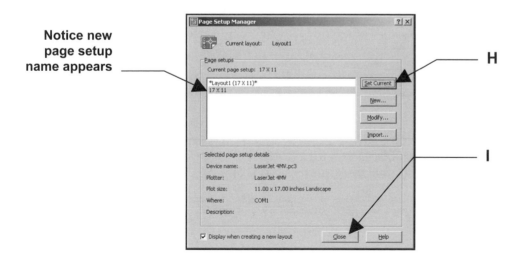

H. Select the **Set Current** button.
I. Select the **Close** button.

You should now have a sheet of paper displayed on the screen.
This sheet is the size you specified in the "Page Setup".
This sheet is in front of the drawing that is in Model Space.
The dashed line represents the printing limits for the device that you selected.

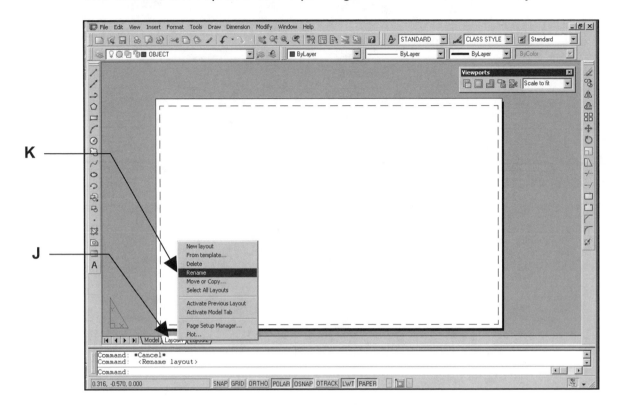

J. Right click on the Layout tab and select **Rename**.
K. Type the new name then select **OK**.

Note: Renaming is optional but helpful. The name is usually a description of the layout you just created. Example: 11 x 17 or Assembly or Floor Plan.

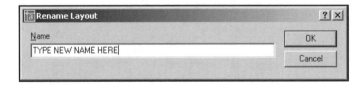

STEP 3. <u>CREATE A BORDER, TITLE BLOCK AND CUT A VIEWPORT.</u>

A. Draw the border, title block and notes in Paper Space.
B. Cut a viewport to see through to Model Space.

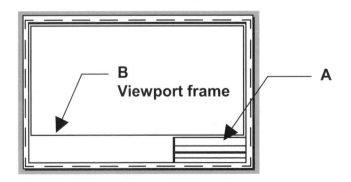

STEP 4. ADJUST MODEL SPACE SCALE and LOCK the VIEWPORT.

Adjust Model Space Scale

A. Double click inside the viewport to change to model space.
    Note: The viewport must be unlocked to adjust the scale.

B. Select **View / Zoom / All** to make sure all of the drawing appears within the
    Viewport.

C. Open the Viewports toolbar. (Refer 2-9)

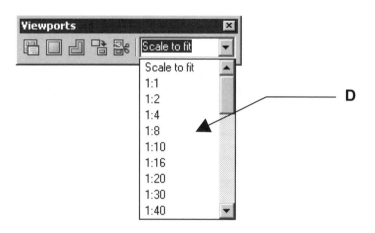

D. Select a scale from the drop down list.

Note: If you need to reposition your drawing within the viewport, use the **PAN** command
(See page 29-2). Do not use the Zoom commands. The zoom commands will change
the scale that you just adjusted.

E. Lock the Viewport

    Lock the Viewport using one of the methods on page 26-11. Now you may
    zoom as much as you desire and it will not affect the adjusted scale.

Adjusting the scale of the viewport will be discussed more in lesson 27.

HOW TO PLOT FROM PAPER SPACE

The previous page setup instructions were to select the printer and paper size. Now you need to specify how you want to plot the drawing. You will find the PLOT dialog box almost identical to the Page Setup dialog box.

STEP 1. SPECIFY PLOT SETTINGS

Specifying Plot Settings means:
Verify the Layout name, Plot device and Paper size.
Select what area of the drawing to plot, what scale to use, where to place the drawing on the paper and which plot style table to use.

A. Select **FILE / PLOT.**

The Plot dialog box shown below should appear.
Select the "More Options" button (in the lower right corner) if your dialog box does not appear the same as shown below.

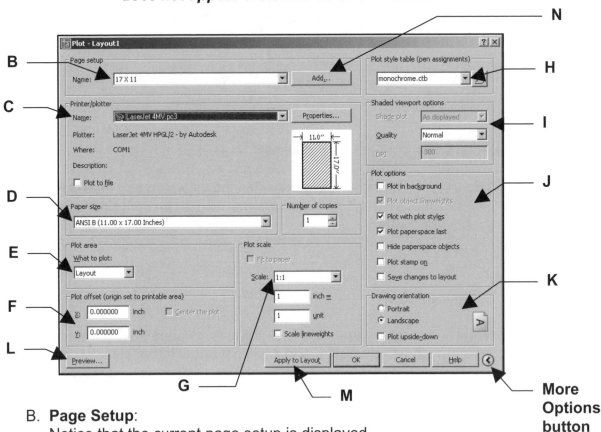

B. **Page Setup**:
Notice that the current page setup is displayed.
You may select a previously saved page setup from the drop down list.
If you make any changes to the settings, this name will change to <none>.

C. **Printer / Plotter:**
Verify the **Name.** All previously configured devices are listed.
If you would like to use your printer, selected it now instead of the Hp4MV shown above.
(If your printer / plotter is not listed, refer to
"Add a Printer / Plotter" Appendix A.)

D. **Paper Size:**
Verify the **Paper size**.

E. **Plot Area:**
Select the area to plot. Layout is the default.
Note: If you selected printer <u>HP4MV</u> select "Layout".
If you selected <u>your printer</u>, use "Extents".

Limits plots the area inside the drawing limits.
(Only shown when plotting from model space)
Layout plots the paper size
(Select when plotting a Layout)
Extents plots all objects in the drawing file even if out of view.
(Only shown if you have a viewport cut)
Display plots the drawing as displayed.
Window plots objects inside a window. To specify the window, choose
Window and designate the first and opposite (diagonal) corner of
the area you choose to plot. (Similar to the Zoom / Window
command)

F. **Plot offset:**
The plot can be moved away from the lower left corner by changing the X
and/or Y offset.
<u>If you select Plot area "Display" or "Extents", select "**Center the plot.**"</u>

G. **Scale:** Select a **scale** from the drop down list or enter a custom scale.
Note: If you selected printer HP4MV, select scale 1 : 1.
If you selected your printer, select "<u>Fit to Paper size</u>".

Note: This scale is the Paper Space scale. The Model space scale was
*adjusted previously within the viewport. If you are plotting from a "**LAYOUT**"*
*tab, normally you will use **plot scale 1:1**. (I know this seems a little confusing*
right now. Scaling will be discussed more in Lesson 27)

H. **Plot Style Table:** Select the Plot Style Table from the list.
(To create your own refer to Appendix C)

I. **Shaded viewport options**
This area is used for printing shaded objects when using 3D and will be
discussed in the Advanced workbook.

J. **Plot options**
Plot Object Lineweights = plots objects with assigned lineweights.
Plot with Plot Styles = plots using the selected Plot Style Table.
Plot paperspace last = plots model space objects before plotting paperspace
objects. Not available when plotting from model space.
Hide Paperspace Objects = used for 3D only. Plots with hidden lines
removed.
Plot Stamp = Allows you to print information around the perimeter of the
border such as; drawing name, layout name, date/time, login name, device
name, paper size and plot scale.
Save Changes to Layout = Select this box if you want to save all of these
settings to the current Layout tab.

K. **Drawing Orientation**.
    Portrait = the short edge of the paper represents the top of the page.
    Landscape = the long edge of the paper represents the top of the page.

L. Select **Preview** button.
    Preview displays the drawing as it will plot on the sheet of paper.

(Note: If you cannot see through to Model space, you have not cut your viewport yet)

If the drawing is centered on the
sheet, press the **Esc** key and
continue.

If the drawing does not look
correct, press the **Esc** key and
check all your settings, then
preview again.

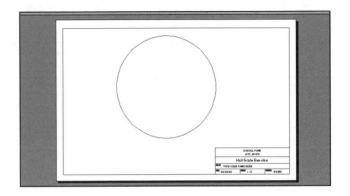

Note: It you set your Viewport layer to "No Plot" the Viewport "frame" will not appear
when you select Preview. What you see is what will print on the paper.

M. **Apply to Layout**
    This applies all of the previous settings to the layout tab. Whenever you
    select this layout tab the settings will be already set for you.

N. At this point you have the option of saving these settings as another page
    setup for future use, not just this layout tab. If you wish to save this setup,
    select the **ADD** button, type a name and select **OK.**

If your computer is connected to the plotter / printer selected, select the **OK**
button to plot, then proceed to **O**.

*If your computer is **not** connected to the plotter / printer selected*, select the
Cancel button to close the Plot dialog box and proceed to **O**.

Note: Selecting Cancel will cancel your selected setting if you did not select the
"Apply to Layout" button or save the page setup as described in **N**.

O. Save the drawing
    This will guarantee that the Page Setup you just created will be saved to this
    file for future use.

Note: This is the concept only.
The step by step instructions start in Exercise 26A

SETTING THE PICK BOX SIZE

☐ Pick box

When AutoCAD prompts you to *select objects,* such as when you are erasing objects, the cursor (crosshairs) turns into a square. This square is called a **Pick box.** The size of the **Pick box** can be changed. Some AutoCAD users prefer large boxes, some like small boxes. The size of the box is your personal preference, however, the smaller the Pick box the more **accurate** you must be when placing the pick box on an object to select it. If the Pick box is too large it could overlap onto other objects that you did not want to select.

HOW TO CHANGE THE "PICK BOX" SIZE.

1. Select **TOOLS / OPTIONS**
2. Select the **SELECTION** tab.

   The following dialog box will appear.

3. Adjust the size by sliding the tab (click and drag) to the right (max) or to the Left (Min.). A preview of the new size is displayed.

2

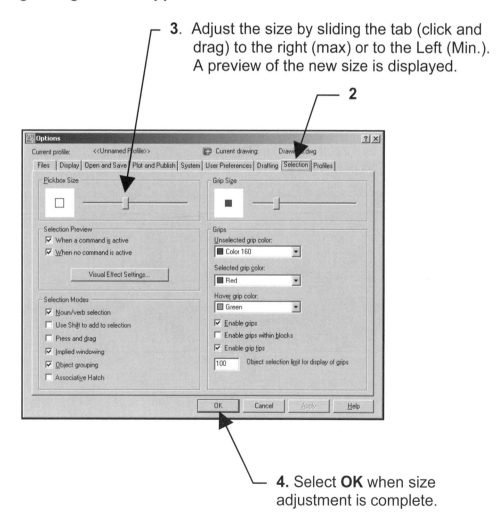

4. Select **OK** when size adjustment is complete.

EXERCISE 26A
CREATE A MASTER DECIMAL SETUP DRAWING

The following instructions will guide you through creating a "Master" decimal setup drawing. The "1Workbook Helper" is an example of a Master setup drawing. Even though the screen appears blank, the actual file is full of settings, such as: Units, Drawing Limits, Snap and Grid settings, Layers, Text styles and Dimension Styles. Once you have created this "Master" drawing, you just open it and draw. No more repetitive inputting of settings.

NEW SETTINGS

A. Begin your drawing without a template as follows:

    1. Select **"FILE / NEW"**.
    2. Select **"START FROM SCRATCH"** Box.
    3. Select "Imperial" or "Metric".
    4. Select **"OK"**.
     Your screen should be blank,
     no grids and the current layer is 0.

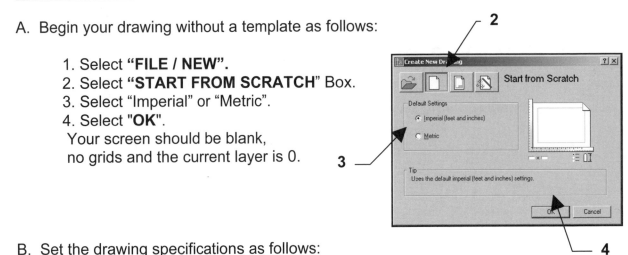

B. Set the drawing specifications as follows:

    1. Select **FORMAT / UNITS.**
      a. Change the "Type" and "Precision" then select **OK.**

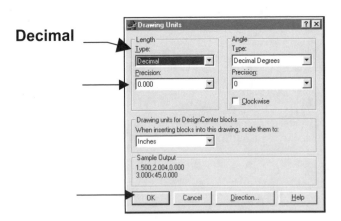

    2. Select **FORMAT / DRAWING LIMITS** (Size of the drawing area)
      a. Lower left corner = 0.000,0.000
      b. Upper right corner = 17 , 11
      c. Use **"VIEW / ZOOM / ALL"** to generate the new limits
      d. Set your **Grids** to **ON** to display the paper size.

3. Select "**TOOLS / DRAFTING SETTINGS / SNAP and GRID** tab
   a. Change the settings as shown below and then select **OK**.

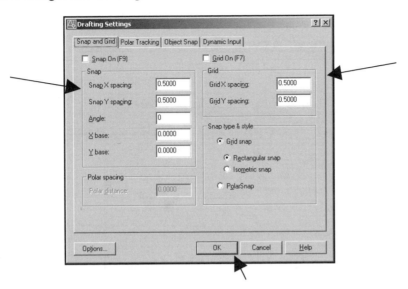

4. Set **"PICK BOX"** size (your preference)
   See page 26-19 for instructions.

NEW LAYERS

C. Create new layers

    1. First **Load** the linetypes listed below. (See page 26-5 for instructions)
       CENTER2
       HIDDEN
       PHANTOM2

    2. Change the Lineweight settings to inches or metric and adjust the Display
       scale.   (See pg. 9-7)

    3. Assign names, colors, linetypes, lineweights and plotability.
       (See page 26-3 for instructions)

NAME	COLOR	LINETYPE	LWT	PLOT
BORDER	RED	CONTINUOUS	.039 (1.00mm)	YES
CENTER	CYAN	CENTER2	Default	YES
CONSTRUCTION	WHITE	CONTINUOUS	Default	NO
DIMENSION	BLUE	CONTINUOUS	Default	YES
HATCH	GREEN	CONTINUOUS	Default	YES
HIDDEN	MAGENTA	HIDDEN	Default	YES
OBJECT	RED	CONTINUOUS	.024 (0.60mm)	YES
PHANTOM	MAGENTA	PHANTOM2	Default	YES
SECTION	WHITE	PHANTOM2	.031 (0.80mm)	YES
TEXT HEAVY	WHITE	CONTINUOUS	Default	YES
TEXT LIGHT	BLUE	CONTINUOUS	Default	YES
THREADS	GREEN	CONTINUOUS	Default	YES
VIEWPORT	GREEN	CONTINUOUS	Default	NO

NEW TEXT STYLE

D. Create a text style

    1. Select "**FORMAT / TEXT STYLE**".
    2. Create a new text style using the settings shown below.

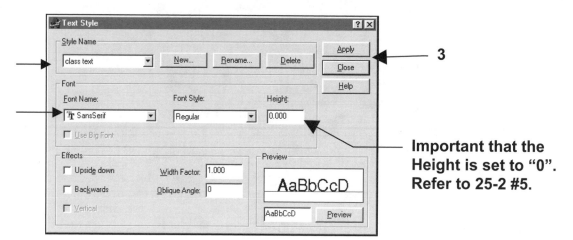

Important that the Height is set to "0". Refer to 25-2 #5.

    3. When complete, select **APPLY** then **CLOSE**.

NEW DIMENSION STYLE

E. Create a new Dimension Style named *Class Style.*
(Refer to Page 16-8 for step-by-step instructions.)

NEW DIMENSION SUB-STYLE

F. Create a new Dimension Sub-Style for *Radius.*
(Refer to Page 18-6 for step by step instructions.)

THIS NEXT STEP IS VERY IMPORTANT.

G. **SAVE ALL THE SETTINGS YOU JUST CREATED.**
    1. Select File / Save as.
    2. Save as: **My Decimal Setup.**

H. Now continue on to Exercise 26B.

EXERCISE 26B
CREATE A BORDER FOR PLOTTING

The following instructions will guide you through creating a Border drawing that will be used in combination with "My Decimal Setup" when plotting. You will create a Layout and draw a border with a title block. All of this information will be saved and you will not have to do this again.

A. Open **My Decimal Setup**

B. Select a **LAYOUT** tab.

Note: If the "Page Setup Manager" dialog box shown below does not appear automatically, right click on the Layout tab, then select Page Setup Manager.

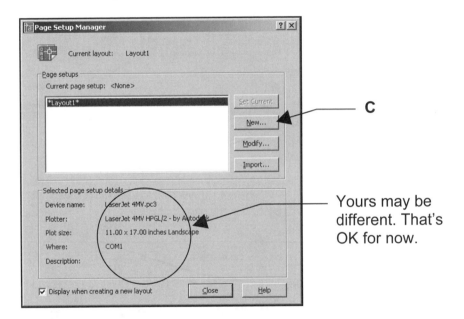

Yours may be different. That's OK for now.

C. Select the **New** button. (*The "New Page Setup" dialog should appear*)

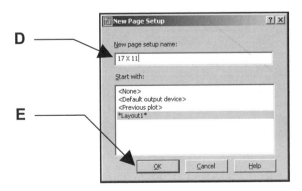

D. Type the new page setup name, **17 x 11**.
E. Select the **OK** button.

(The "Page Setup" dialog box should appear)

If this printer
does not appear
in the list, refer
to Appendix A.

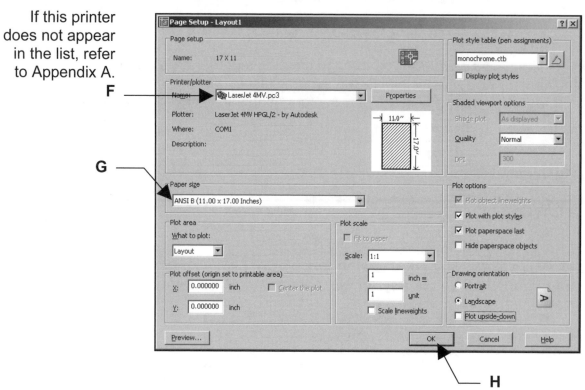

F. Select the **Printer / Plotter**.
G. Select the **Paper Size.**
H. Select the **OK** button.

**The "Page Setup Manager" should have returned. Notice the new page
setup name now appears in the list.**

Notice new
page setup
name appears

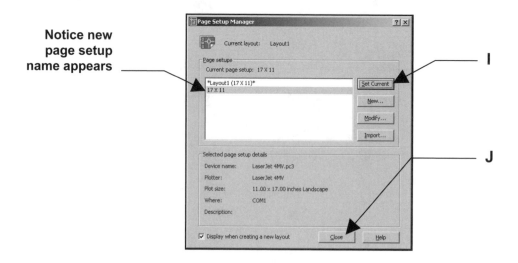

I. Select the **Set Current** button. (The paper should have changed)
J. Select the **Close** button.

You should now have a sheet of paper displayed on the screen.
This sheet is the size you specified in the "Page Setup".
This sheet is in front of Model Space.
The dashed line represents the printing limits for the device that you selected.

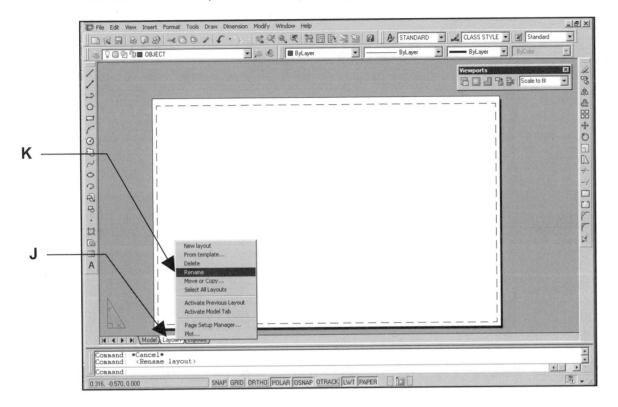

J. Right click on the Layout tab and select **Rename**.
K. Type the new name **11 X 17 (1 to 1)** then select **OK**.

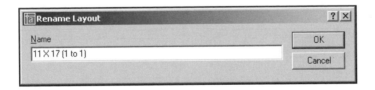

Note: The display color of the modelspace and paperspace may be changed using
TOOLS / OPTIONS / DISPLAY / COLORS. This is discussed in the Advanced
workbook.

L. Draw the Border with title block, shown below, on the sheet of paper shown on the screen.

M. When you have completed the Border, shown below:
1. Select File / Save as
2. Save as: **My Decimal Setup** (Again)

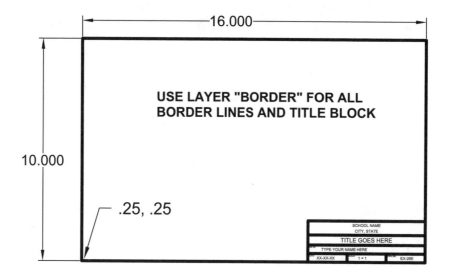

Important: Use "Single Line Text" to place the text in the Title Block below.

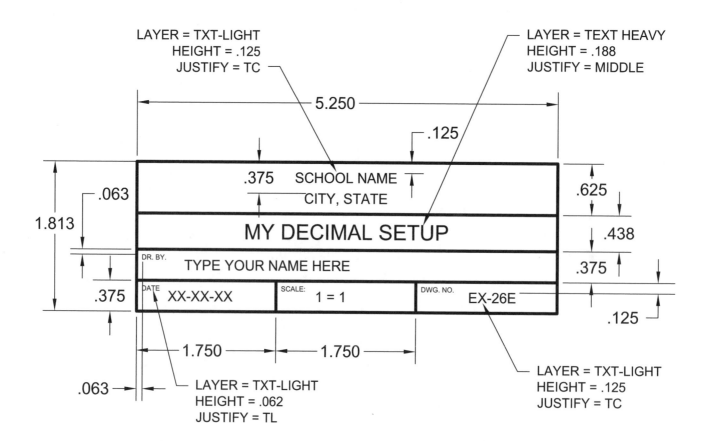

EXERCISE 26C
CREATE A VIEWPORT

The following instructions will guide you through creating a VIEWPORT in the Border Layout sheet. Creating a viewport has the same effect as cutting a hole in the sheet of paper. You will be able to see through the viewport frame (hole) to Modelspace.

A. Open **My Decimal Setup**
B. Select the **11 X 17 (1 to 1)** tab.
C. Select layer "Viewport"
D. Open the "Viewports" toolbar (page 2-9) and select "Single" viewport icon.
E. Draw a Single viewport approximately as shown.

Single Viewport icon (LT's toolbar looks a little different but it works the same)

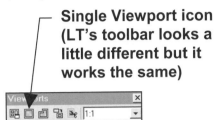

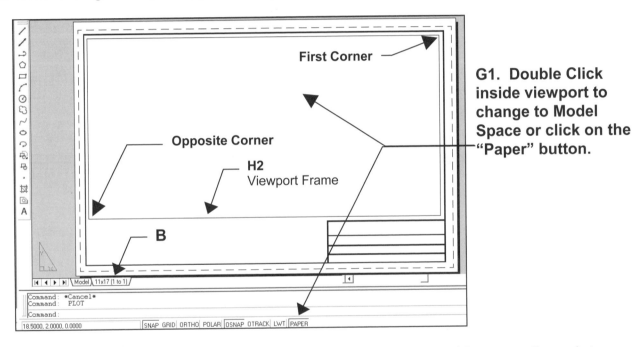

First Corner

Opposite Corner

H2
Viewport Frame

B

G1. Double Click inside viewport to change to Model Space or click on the "Paper" button.

F. After successfully creating the Viewport, you should now be able to see through to Modelspace. (Your grids should appear <u>if they are ON</u>.)

G. Adjust the Model space scale.
   1. Select the **Paper** button so it changes to "Model" or double click inside the Viewport Frame.
   2. Do **Zoom / All** before adjusting the scale.
   3. Select 1:1 in the **VIEWPORT** toolbar.

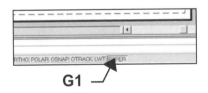

G1

H. Lock the Viewport
   Refer to 26-11 for Locking instructions.

G2

J. Save as: **My Decimal Setup**

K. Continue on to 26-D.

EXERCISE 26D
PLOTTING FROM PAPERSPACE

The following instructions will guide you through the final steps for setting up the master file for plotting. These settings will stay with **My Decimal Setup** and you will be able to use it over and over again. (Refer to Page 26-16 for more detailed explanations)

A. Open **My Decimal Setup** if it isn't already open.

B. Select the **11x17 (1 to 1)** layout tab.

   You should be looking at your Border and Title Block now.

C. Select **File / Plot** or place the cursor on the **11x17 (1 to 1)** tab and press the right mouse button. Select **Plot** from the short cut menu.

The Plot dialog box shown below should appear.
Select the "More Options" button if your dialog box does not appear the same as shown below

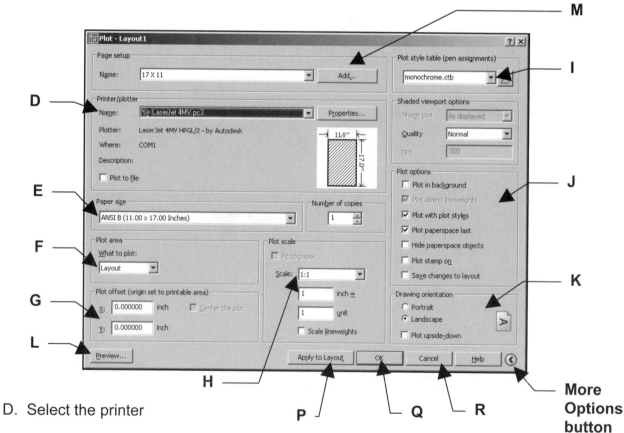

D. Select the printer

E. Select the Paper Size

F. Select the Plot Area

G. Plot offset should be 0 for both X and Y.

H. Select scale 1 : 1

I. Select the Plot Style Table "Monochrome.ctb"

J. Select the Plot options shown

K. Select Drawing Orientation: Landscape

L. Select **Preview** button.

    1. If the drawing is centered on the sheet, press the Esc key and continue on to **M**.

    2. If the drawing does not look correct, press the Esc key and check all your settings, then preview again.

M. Select the **ADD** button.

N. Type the new page setup name: 11 X 17 Mono

O. Select OK button.

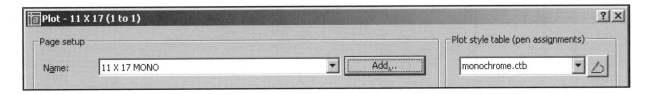

The settings are now saved for future use.

P. Select the "Apply to Layout" button.

Q. If your computer **is** connected to the plotter / printer selected, select the **OK** button to plot, then proceed to **S**.

R. If your computer is <u>not</u> connected to the plotter / printer selected, select the Cancel button to close the Plot dialog box and proceed to **S**. Note: Selecting Cancel <u>will cancel</u> your selected setting if you did not save the page setup as specified in **M**.

S. Save the drawing
    1. Select File / Save As
    2. Save as: **My Decimal Setup**

*You have now completed the **Page Setup** for the **My Decimal Setup**.   Now you are ready to use this master file to create many drawings in the future. <u>In fact, you have one on the very next page.</u>*

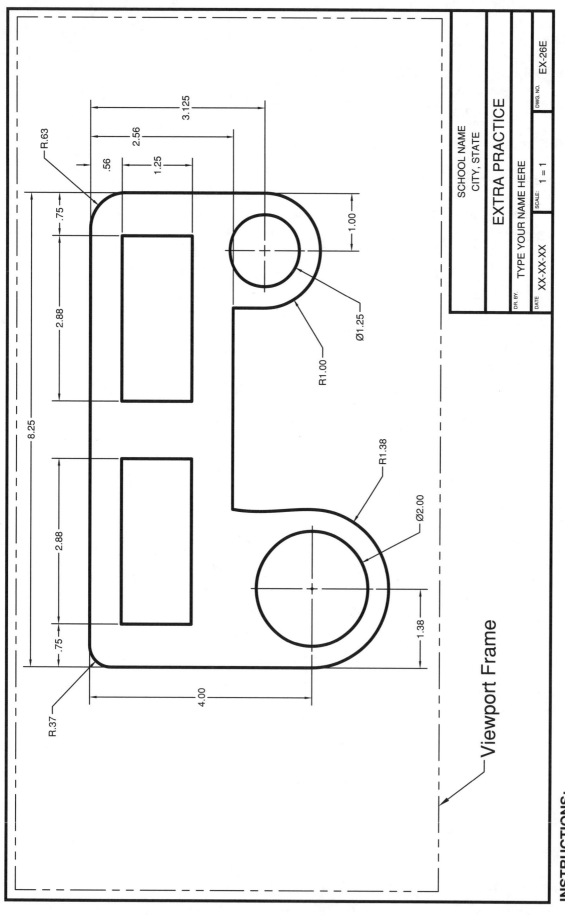

Viewport Frame

EXERCISE 26E

INSTRUCTIONS:

1. Open **MY DECIMAL SETUP**
2. Draw the drawing above inside the viewport frame (Model Space).
3. Check to make sure the viewport's scale is 1:1 and locked. (Refer to EX-26C)
4. Dimension as shown, in paperspace. Use Dimension Style, "Class Style".
5. Save as: EX-26E
6. Plot using "11 X17 MONO" Page Setup.

LEARNING OBJECTIVES

After completing this lesson, you will be able to:

1. Understand scaled drawings.
2. Adjust the scale within a viewport.
3. Calculate the Drawing Scale Factor.
4. Understand how scale affects Text, Hatch and Dimensions.
5. Dimension a scaled drawing.
6. Create an Feet-Inches "Setup" file for future use.
7. Create a new Architectural Border to use when plotting.

LESSON 27

CREATING SCALED DRAWINGS

In the lessons previous to Lesson 26 you worked only in Model space. Then in Lesson 26 you learned that AutoCAD actually has another environment called Paper Space, or Layout. In this lesson we need to learn more about why we need 2 environments and how they make plotting drawings easier.

A very important rule in CAD you must understand is: "All objects are drawn full size". In other words, if you want to draw a line 20 feet long, you actually draw it 20 feet long. If the line is 1/8" long, you actually draw it 1/8" long.

Drawing and Plotting objects that are very large or very small.

In the previous lessons you created medium sized drawings. Not too big, not too small. But what if you wanted to draw a house? Could you print it to scale on a 17 X 11 piece of paper? How about a small paper clip. Could you make it big enough to dimension? Let's start with the house.

Drawing something <u>large</u> such as a house.

1. Start a new drawing from scratch. (26-20)
2. Set the units for the drawing to Architectural.
3. Set the snap to 3 inches and set the grids to 1 foot.
4. Set the drawing limits, in model space, to:
   Lower left corner: 0, 0 Upper Right Corner: 45', 35' (feet), then (Zoom / All).
   Now your drawing area is big enough for the entire house to be drawn <u>full size</u>.
5. Draw a rectangle: 30' L X 20' W (representing the house) and draw a pretend roof.

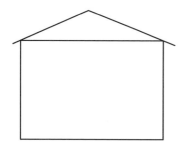

Plotting the house drawing

Now you want to plot this house drawing on a 17 x 11 piece of paper. This is where paper space makes it easy.

6. Select the Layout tab (paper space).
7. When the Page Setup Manager appears, select NEW, give it a name, then OK.
8. Specify the plotter LaserJet 4MV, the paper size 11 x 17 then OK.
9. Select "Set Current" and "Close".
10 Open the "Viewport" tool bar and cut a viewport so you can see through the Viewport to the house drawing that is in model space. (Refer to 26-8)
11 Double click inside the Viewport to reach through to model space. (26-10)
12 Select <u>View / Zoom / Extents</u> so the entire house is visible within the viewport frame.

Now here is where the magic happens.

13. Adjust the scale of model space. (See 27-5)
    Scroll down the list of scales and select ¼" = 1'.

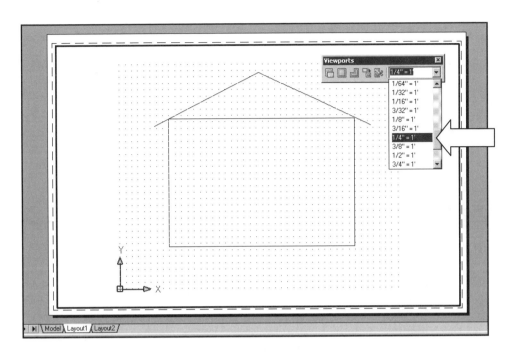

Wow! A 30ft x 20ft house fit on to an 11" x 17" sheet of paper.

Let's talk about what really happened.
Remember the photo and picture frame example I suggested in lesson 26? (Refer to 26-7) This time try to picture yourself standing in front of your house with an empty picture frame in your hands. Look at your house through the picture frame. The house is way too big to fit in the frame. So you walk across the street and look through the picture frame again. Does the house appear smaller? If you could walk far enough away from the house it would eventually appear small enough to fit in the picture frame. But....the house did not actually change size, did it? It only appears smaller because you are farther away from it.

This is the concept that you need to understand.

When you adjust the scale of model space it does not change the actual size of the objects. They just appear smaller because they are farther away from paper space. When you dimension the house, the dimension values will be the actual measurement of the house. In other words, the 30 ft. width will have a dimension of 30'-0".

14. Lock the viewport. (Refer to 26-11)
    When you have adjusted the model space scale to your satisfaction you should "lock" the viewport so the scale will not change when you zoom.
15. Save this drawing as "HOUSE".
16. Plot the paper space / model space combination using File / Plot.
    The Page Setup scale should remain 1 : 1 (Remember you have already adjusted the model space scale. You do not need to scale the combination it is just fine.)
17. Don't forget to Preview before plotting.

Drawing something <u>small</u> such as a paperclip.

When plotting something smaller you have to move the picture frame closer to model space rather than farther away. So in that case, you would adjust the model space scale to a scale larger than 1 : 1 until the object in model space appears large enough to see easily and dimension. Remember, even though the object appears larger, when you dimension it the dimension values will be the correct information.

Let's try a small object.

1. Open **My Decimal Setup.**
2. Select the "Model" tab.

3. Draw a Rectangle 2" L X 1" W. (We won't actually draw the paperclip right now.)

4. Select the "**11 x 17 (1 to 1)** " tab.

You should now be able to see the Rectangle in the viewport.

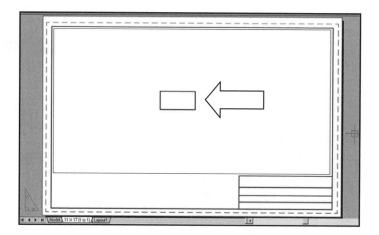

5. Unlock the Viewport.
6. Double click inside the Viewport to get into model space.
7. Adjust the scale of model space to 4 : 1 .

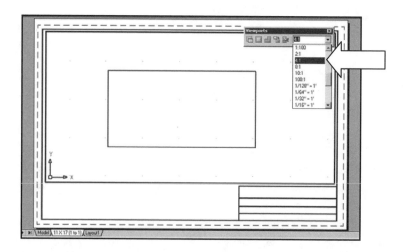

Now the Rectangle is large enough to see clearly and dimension.
You will understand this concept better after completing the exercises in this lesson.

ADJUSTING THE SCALE INSIDE A VIEWPORT

The following will take you through the process of adjusting the scale within a viewport.

NOTE: *DO NOT ADJUST THE SCALE OF "PAPER SPACE".*
PAPER SPACE SCALE SHOULD REMAIN 1 : 1.

1. **Open** a drawing.

2. Select a **Layout** tab. (paper space)

3. Unlock the Viewport if it is locked. (Viewport must be unlocked to adjust the scale)

4. Double click inside of the Viewport to change to Model Space.

   It is very important that you be inside the Viewport.
   You want to adjust the scale of the Model geometry not the Layout – Paper
   Space. Paperspace remains 1:1.

5. Select **View / Zoom / All** before adjusting the scale of the viewport.

6. *Adjust the scale* of the Model geometry by selecting a scale from the
   "Viewports" toolbar.

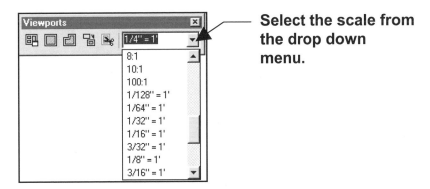

Select the scale from the drop down menu.

(There are 32 preset decimal and architectural scales. Instructions to add or delete scales in this list is shown in the Advanced workbook.)

If you would like a scale other than one listed in the drop down list you may use the Zoom XP command as follows:

a. Select **View / Zoom / Scale.**
b. Enter a scale factor (nX or nXP): ***1/48xp*** *(XP means "times Paperspace".)*

The Drawing Scale Factor for 2 : 1 is 1/2, enter the reciprocal: 2xp.
The Drawing Scale Factor for 1/4"= 1' is 48, enter the reciprocal: **1/48xp**.

(Refer to page 27-6 for "How to calculate the drawing scale factor".)

CALCULATING THE DRAWING SCALE FACTOR

Think about this….
When you adjust the scale within a Viewport all objects within the Viewport appear either larger or smaller. ALL OBJECTS, including <u>Text</u>, <u>Hatch</u> and <u>Dimension</u> features such as arrowheads. Even the spacing for non-continuous linetypes such as "dashed" appear larger or smaller. As a result, these objects may be too small or too large. So you must adjust the size of these objects.

If the objects appear smaller, you must increase the size of the Text height, Hatch scale or spacing and Dimension entities. If the objects appear larger, you must decrease the size of the Text height, Hatch scale or spacing and Dimension entities. I guess you figured that much out on your own.

But how do you know what size to make them?

Determining the size
To determine how much to increase or decrease you need to calculate the **<u>drawing scale factor</u>**. The drawing scale factor (DSF) means "<u>How many times smaller or larger did the drawing get when you adjusted the scale</u>".

Here are a few of the most commonly used scales and their <u>drawing scale factors</u>:

Scale	DSF	Scale	DSF
1" = 1'	12	1 : 2	2
1/4" = 1'	48	2 : 1	1/2
1/8" = 1'	96	4 : 1	1/4

<u>HOW TO CALCULATE THE DRAWING SCALE FACTOR (DSF)</u>:

The drawing scale factor is the reciprocal of the adjusted scale.

For example, if the scale has been adjusted to: 1/4" = 1'
calculate the scale factor as follows:
1. Adjusted scale: ¼" = 1'
2. Convert to decimals: .25 = 12
3. Divide 12 by .25 (12 ÷ .25 = 48)   (The Drawing scale factor is 48)

This means that the text height, hatch scale/spacing and dimension scale must be adjusted by 48. (Refer to more detailed instructions on the following pages)

Another example: if the scale has been adjusted to 4 : 1
1. Adjusted scale: 4 : 1
2. Convert to decimals: 4 = 1
   (if necessary)
3. Divide 1 by 4   (1 ÷ 4 = 1/4)   (The Drawing scale factor is 1/4 or .25)

This means that text height, hatch scale/spacing and dimension scale must be adjusted by 1/4. (Refer to more detailed instructions on the following pages)

HOW THE DSF EFFECTS TEXTS

When you adjust the scale within a Viewport, any text within the Viewport (in model space) will appear either larger or smaller, depending on the adjusted scale.

To calculate the correct text height, do the following:
1. Decide what height you would like the text to be when plotted on the paper.
2. Calculate the Drawing Scale Factor as follows:
   Multiply the "Drawing Scale Factor" (refer to 27-6) times the "Plotted Text height".

For example:
If you want the text height after plotting to be **1/8",** and the
adjusted scale within the viewport is 4 : 1 (<u>DSF calculated = **1/4**</u>)
Multiply 1/8 (Plotted Text Height) X 1/4 (DSF) = **1/32"**

So the text height you use is 1/32". Now I know that seems like it will be very small, but remember, the viewport scale has been adjusted to 4 : 1 (4X it's original appearance. So if you set a text height of 1/32" it will be enlarged 4X and have an appearance of 1/8". (That is the height you want when it is plotted)

The following is an example of how the text will appear when plotted.

WRONG: This is what the text would look like if you used 1/8" ht. It appears too large because the model space scale has been adjusted to 4 : 1 (4 times larger)

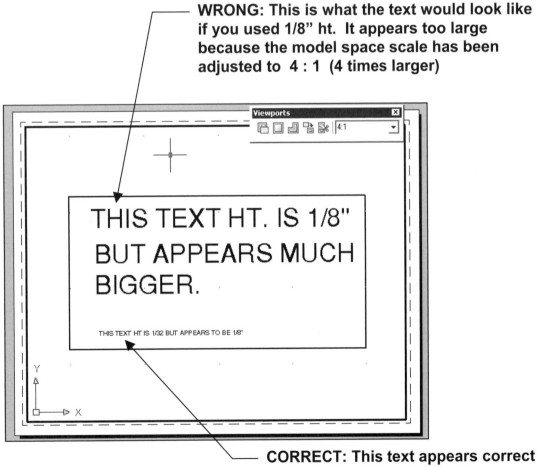

THIS TEXT HT. IS 1/8"
BUT APPEARS MUCH
BIGGER.

THIS TEXT HT IS 1/32 BUT APPEARS TO BE 1/8"

**CORRECT: This text appears correct because you took into consideration the adjusted scale.
The text is 1/32 ht times 4 = 1/8**

HOW THE DSF EFFECTS HATCH

When you adjust the scale within a Viewport the spacing between the Hatch lines will appear either larger or smaller depending on the scale.

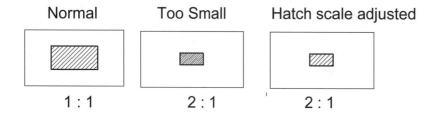

Normal	Too Small	Hatch scale adjusted
1 : 1	2 : 1	2 : 1

PREDEFINED

Calculate the Drawing Scale Factor and enter the DSF in the **Scale** box.

Enter DSF here.

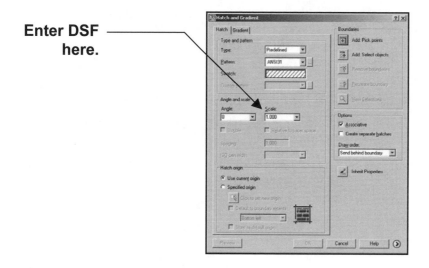

USER DEFINED

Calculate the Drawing Scale Factor and enter the DSF in the **SPACING** box.

Enter DSF here

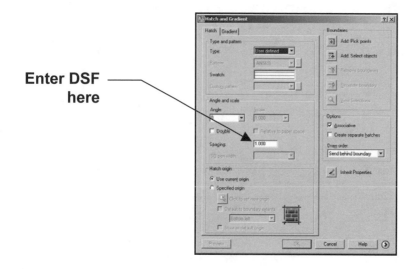

HOW THE DSF EFFECTS DIMENSIONING

Note: the following only affects dimensions in model space.
Refer to 26-10 for dimensions in paper space.

When you adjust the scale within a Viewport, dimension features, such as arrowheads, will increase or decrease in appearance also.

Do not misunderstand. It will not affect the dimension "value", only the size of the dimension features, such as arrowheads.

All you need to do is set the Dimension "overall scale" box to the Drawing Scale Factor. Dimension feature sizes should remain as you normally have them set. Then AutoCAD will automatically adjust the size of all the dimension features as you draw the dimensions in the viewport.

HOW TO SET THE "OVERALL SCALE".

1. Select "**Format / Dimension Style**"
2. Select the dimension style that you want to change.
3. Select the "**Modify**" button.
4. Select the "**Fit**" tab.
5. Select "**Use overall scale of**:"
6. Enter the **DSF** in the box.
7. Select **OK** and close the Dimension Style dialog box.

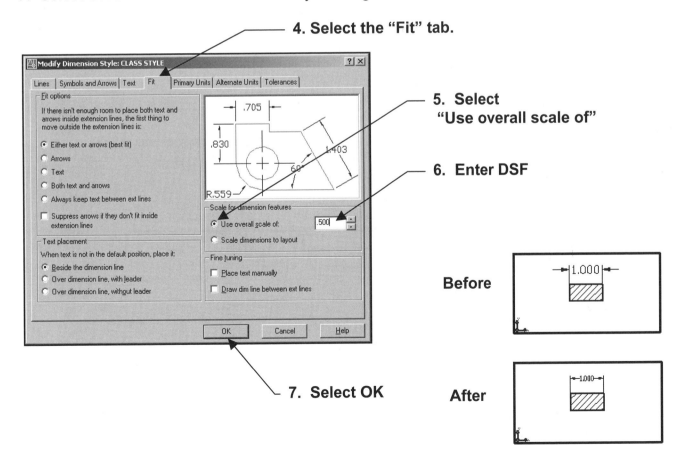

4. Select the "Fit" tab.

5. Select "Use overall scale of"

6. Enter DSF

7. Select OK

Before

After

TRANS-SPATIAL DIMENSIONING

In Lesson 16 you learned about True Associative dimensioning. (Refer to 16-2) Remember, when using True Associative dimensioning, the dimension is actually attached to the object. If the object changes, the dimension changes. True Associative dimensioning is a very important and powerful tool within AutoCAD.

True Associative dimensioning gets even better because it can also be trans-spatial. Trans-spatial means that you have the ability to place a dimension in paper space while the object you are dimensioning is in model space. Even though the dimension is in paper space, it is actually attached to the object, in model space.

For example,
1. You draw a house in model space.
2. Now select the layout tab.
3. Cut a viewport so you can see the drawing of the house.
4. Go to model space, adjust the scale of the viewport (model space) and lock it.
5. Now go to paper space and dimension the house.

Why is Trans-spatial dimensioning so great?
If you dimension in Paper Space you do not have to be concerned with the Drawing Scale Factor. The overall scale described on page 27-9 is set to 1.

SHOULD YOU DIMENSION IN PAPER SPACE OR MODEL SPACE?
It is possible to dimension in either Paper Space or Model Space. Personally, I dimension in either space depending on the situation. Here are some things to consider.

Paper Space dimensioning
Pro's
1. You never have to worry about the drawing scale factor. [Overall scale (page 27-9) will remain set to 1.]
2. Dimensions do not change appearance if you adjust the scale of the viewport.
3. All dimensions will have the same appearance in all viewports.

Con's
1. Qdim is not trans-spatial. (You can only use it in model space.)
2. Dimensions sometimes temporarily float away from objects if you move the objects. (Use "Dimregen" to put them back in place.)

Model Space dimensioning
Pro's
1. Dimensions never float away from objects.
2. Qdim works in modelspace.

Con's
1. You must change the overall scale (page 27-9) to match the drawing scale factor.
2. Dimensions will appear different in each viewport if the viewports have different adjusted scales.

LINETYPE SCALE

AutoCAD has many Linetypes, as you learned in Lesson 26. (26-4) A Linetype is a series of dashes, lines and spaces. Each linetype has specific dimensions for the lines, dashes and spaces. When you are in paper space and you adjust the scale of the model space, AutoCAD automatically adjusts the scale of the linetype dashes, lines and spaces. So you do not have to do any modifications unless you desire.

Changing individual objects
If you want to change the linetype scale for an individual object use the Properties Palette.

Changing the entire drawing
If you want to change the linetype scale for the entire drawing:
1. On the command line type: **LTS <enter>**
2. LTSCALE Enter new linetype scale factor <1.0000>: **enter the new value**

NOTE:

If you are in Paperspace you will only need to change the value to 2 or .5 etc. AutoCAD attempts to adjust the linetype scale automatically to the drawing scale factor. But sometimes you want to tweek it a little.

Linetypes may not appear the same if you are in the model tab rather than the Layout tab. Linetype scale is handled differently in Model space.

Linetype scaling is a visual preference. There is no rule for spacing in cad.

PSLTSCALE
This variable controls paper space linetype scaling. This means that all linetype dash lengths and spaces will be scaled to the paper space scale.

1 = the viewports can have different adjusted scales and the linetype scale will be the same in all viewports. (this is the default setting)

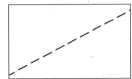

0 = viewports with different adjusted scales will also appear to have differing linetype scales.

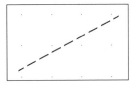

Note: Select View/Regen after changing Variable.

EXERCISE 27A
CREATE A MASTER FEET-INCHES SETUP DRAWING

The following instructions will guide you through creating a "Master" feet-inches setup drawing. The "1Workbook Helper" is an example of a Master setup drawing. Even though the screen appears blank, the actual file is full of settings, such as, Units, Drawing Limits, Snap and Grid settings, Layers, Text styles and Dimension Styles. Once you have created this "Master" drawing you just open it and draw. No more repetitive inputting of settings.

NEW SETTINGS

A. Begin your drawing without a template as follows:

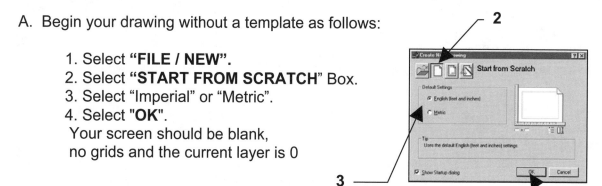

1. Select **"FILE / NEW"**.
2. Select **"START FROM SCRATCH"** Box.
3. Select "Imperial" or "Metric".
4. Select **"OK"**.
   Your screen should be blank,
   no grids and the current layer is 0

B. Set drawing specifications as follows:

1. Select **FORMAT / UNITS**
   a. Change the "Type" and "Precision" then select **OK**.

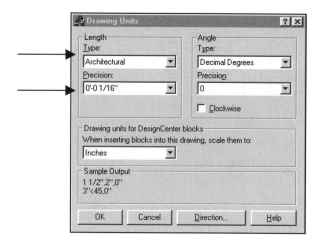

2. Select **FORMAT / DRAWING LIMITS** (Size of the drawing area)
   a. Lower left corner = 0'-0", 0'-0"
   b. Upper right corner = 68', 44'. (Notice this is feet, not inches.)
   c. Use **"VIEW / ZOOM / ALL"** to generate the new limits.

3. Select **TOOLS / DRAFTING SETTINGS / SNAP and GRID** tab.
   a. Change the settings as shown below and then select **OK.**
   b. Set your **grids** to **ON** to display the paper size.

Snap 3"
(inches)

Grid 1' (foot)

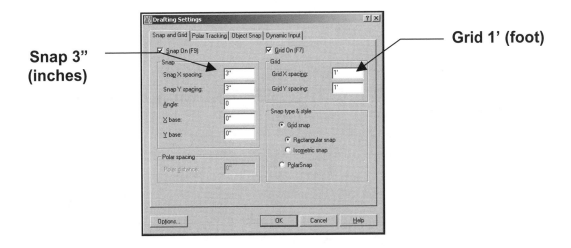

4. Set **"PICK BOX"** size (your preference).
   (See page 26-19 for instructions.)

NEW LAYERS

C. Create new layers.

1. **Load** linetype, (See page 26-5 for instructions.)
   DASHED
2. Change the Lineweight settings to inches and adjust the Display scale.
   (See pg. 9-7)

3. Assign names, colors, linetypes, lineweights and plotability.
   (See page 26-3 for instructions.)

NAME	COLOR	LINETYPE	LWT	PLOT
BORDER	RED	CONTINUOUS	.039 (1.00mm)	YES
CABINETS	CYAN	CONTINUOUS	.016 (0.40mm)	YES
CONSTRUCTION	WHITE	CONTINUOUS	Default	NO
DIMENSION	BLUE	CONTINUOUS	Default	YES
DOORS	GREEN	CONTINUOUS	.016 (0.40mm)	YES
ELECTRICAL	CYAN	CONTINUOUS	.010 (0.25mm)	YES
FURNITURE	MAGENTA	CONTINUOUS	.016 (0.40mm)	YES
HARDSCAPE	WHITE	CONTINUOUS	.024 (0.60mm)	YES
LANDSCAPE	GREEN	CONTINUOUS	.016 (0.40mm)	YES
PLUMBING	9	CONTINUOUS	.010 (0.25mm)	YES
TEXT HEAVY	WHITE	CONTINUOUS	Default	YES
TEXT LIGHT	BLUE	CONTINUOUS	Default	YES
VIEWPORT	GREEN	CONTINUOUS	Default	NO
WALLS	RED	CONTINUOUS	.024 (0.60mm)	YES
WINDOWS	GREEN	CONTINUOUS	.016 (0.40mm)	YES
WIRING	CYAN	DASHED	.010 (0.25mm)	YES

NEW TEXT STYLE

D. Create 2 text styles named **CLASS TEXT** and **ARCH TEXT**

    1. Select "**FORMAT / TEXT STYLE**".
    2. Make the changes shown in the dialog boxes below.

CLASS TEXT   **ARCH TEXT**

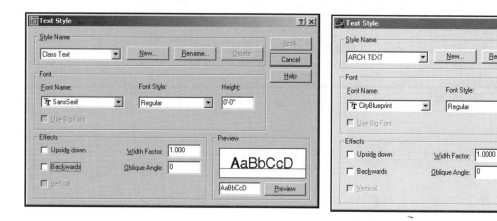

    3. When complete, select **APPLY** then **CLOSE**.

NEW DIMENSION STYLE

E. A new dimension style should also be created, but that will be discussed separately in Exercise 27C, page 27-19.

THIS NEXT STEP IS VERY IMPORTANT.

F. SAVE ALL THE SETTINGS YOU JUST CREATED
    1. Select File / Save as.
    2. Save as: **My Feet-Inches Setup.**

G. Now continue on to Exercise 27B.

EXERCISE 27B
CREATE AN ARCHITECTURAL BORDER FOR PLOTTING

The following instructions will guide you through creating a Border drawing that will be used in combination with "My Feet-Inches Setup" when plotting. You will create a Layout and draw a border with a title block. All of this information will be saved and you will not have to do this again.

A. Open **My Feet-Inches Setup**

B. Select a **LAYOUT** tab.  B

Note: If the "Page Setup Manager" dialog box shown below does not appear automatically, right click on the Layout tab, then select Page Setup Manager.

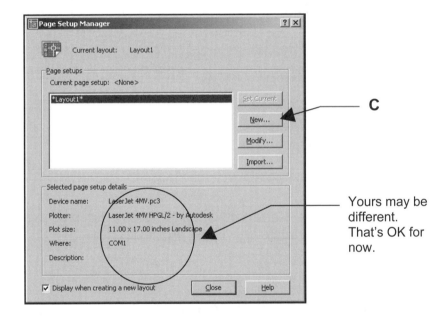

C

Yours may be different. That's OK for now.

C. Select the **New** button. *The "New Page Setup" dialog should appear.*

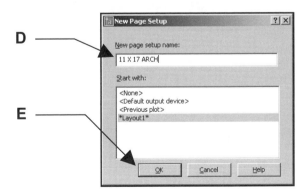

D

E

D. Type the new page setup name, **11 x 17 Arch**.
E. Select the **OK** button.

The "Page Setup" dialog box should appear.

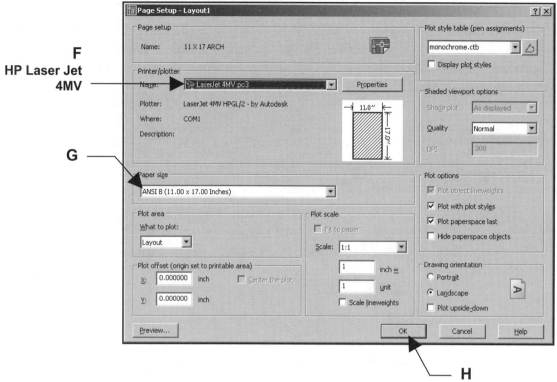

F
HP Laser Jet
4MV

G

H

F. Select the **Printer / Plotter** shown.
G. Select the **Paper Size** shown.
H. Select the **OK** button.

The "Page Setup Manager" should have returned. Notice the new page setup name now appears in the list.

**Notice new
page setup
name appears**

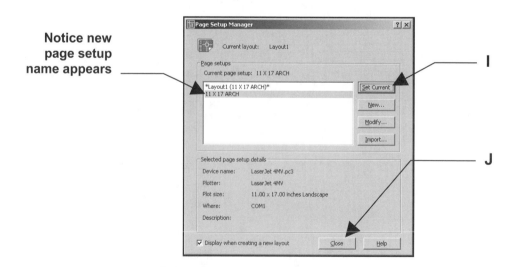

I

J

I. Select the **Set Current** button.
J. Select the **Close** button.

You should now have a sheet of paper displayed on the screen.
This sheet is the size you specified in the "Page Setup".
This sheet is in front of the drawing that is in Model Space.
The dashed line represents the printing limits for the device that you selected.

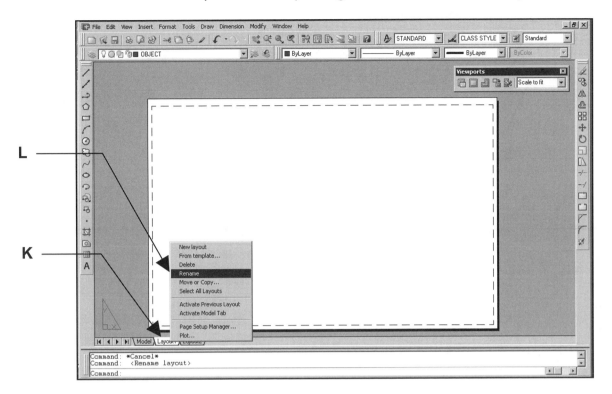

K. Right click on the Layout tab and select **Rename**.
L. Type the new name **11 X 17 Arch** then select **OK**.

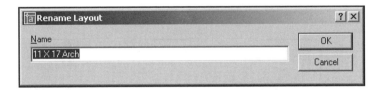

M. Draw the Border with title block, shown below, on the sheet of paper shown on the screen.

N. When you have completed the Border, shown below:
1. Select File / Save as.
2. Save as: **My Feet-Inches Setup** (again).

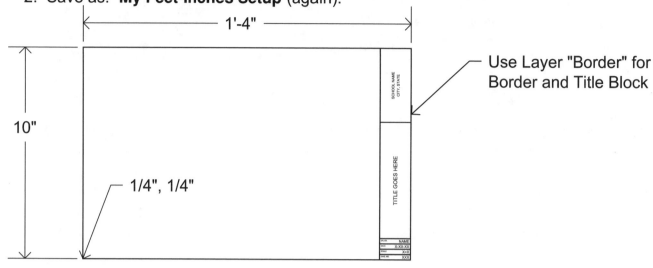

Use Layer "Border" for Border and Title Block

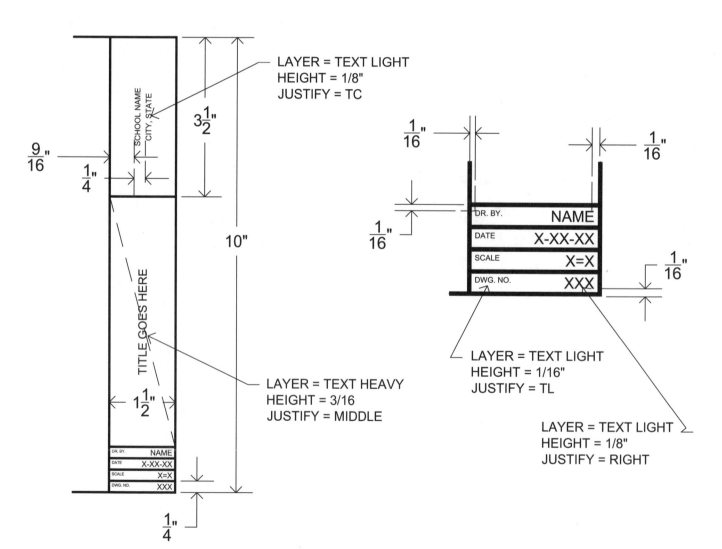

EXERCISE 27C
CREATE A NEW ARCHITECTURAL DIMENSION STYLE

1. Open **My Feet-Inches Setup.**

2. Set **DIMASSOC** to **2**

3. Select **FORMAT / DIMENSION STYLE**

4. Select the **NEW** button.

5. New Style Name: **ARCH DIM**

6. Start With: **Standard**

7. Use For: **All dimensions**

8. Select the **CONTINUE** box.

9. Select the **"Primary Units"** tab and make the following changes.

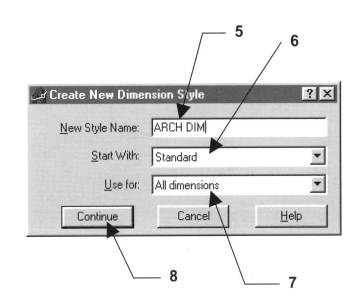

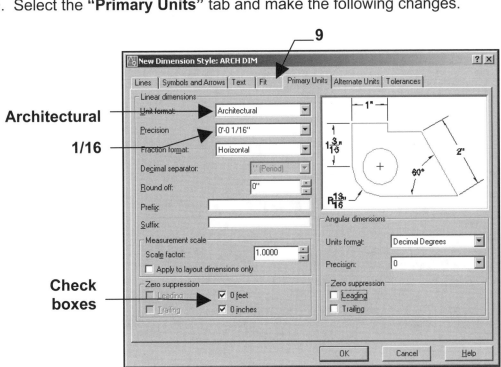

DO NOT SELECT THE OK BUTTON YET

10. Select the *"Lines"* tab and make the following changes.

10

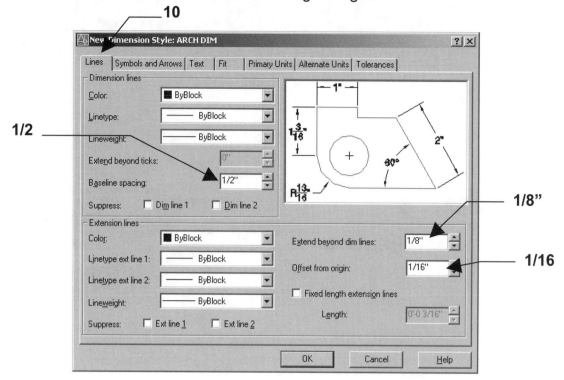

1/2

1/8"

1/16

DO NOT SELECT THE OK BUTTON.

11. Select the **"Symbols and Arrows"** tab and make the following changes.

11

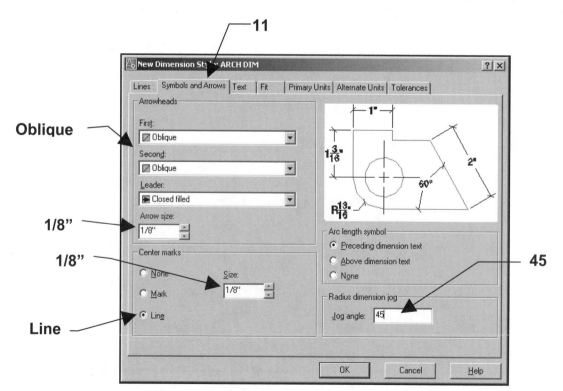

Oblique

1/8"

1/8"

Line

45

DO NOT SELECT THE OK BUTTON

12. Select the **"Text"** tab and make the following changes.

12

Arch Text ────→

1/8" ────→

Above ────→

1/16" ────→

Aligned ────→

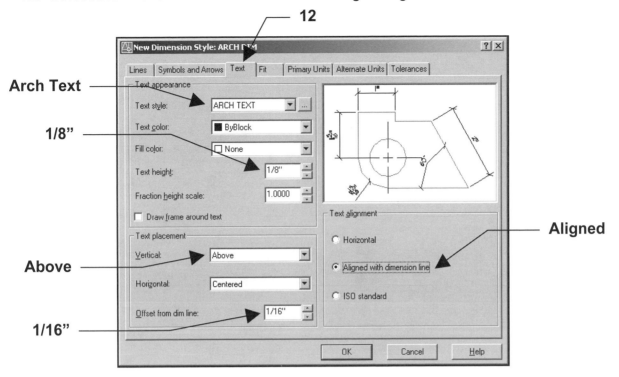

DO NOT SELECT THE OK BUTTON

13. Select the **"Fit"** tab and make the following changes.

13

Check box ────→

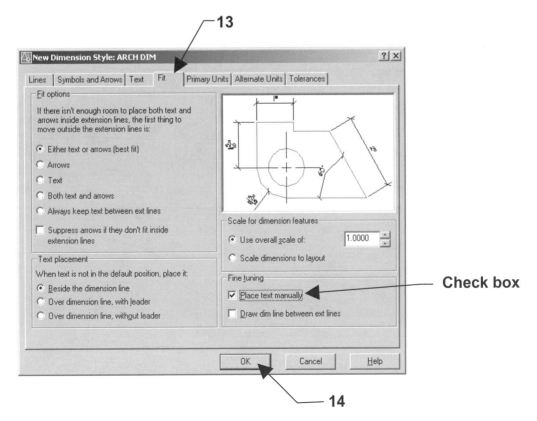

14

14. **NOW** select the **OK** button.

15. *Your new style "__ARCH DIM__" should be listed.*

16. Select the **"Set Current"** button to make your new style "**ARCH DIM**" the style that you will use.

15 —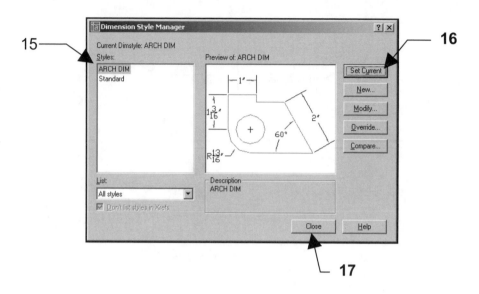

16

17

17. Select the **Close** button to **exit**.

18. Save your drawing as **My Feet-Inches Setup**

19. Continue on to Exercise 27D.

EXERCISE 27D
CREATE A VIEWPORT

The following instructions will guide you through creating a VIEWPORT in the Border Layout sheet. Creating a viewport has the same effect as cutting a hole in the sheet of paper. You will be able to see through the viewport frame (hole) to Model.

A. Open **My Feet-Inches Setup.**
B. Select the **11 X 17 Arch** tab.
C. Select layer "Viewport".
D. Open the "Viewports" toolbar (page 2-9) and select "Single" viewport icon.
E. Draw a Single viewport approximately as shown.

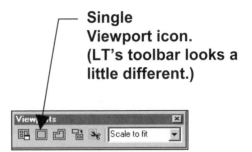

Single Viewport icon. (LT's toolbar looks a little different.)

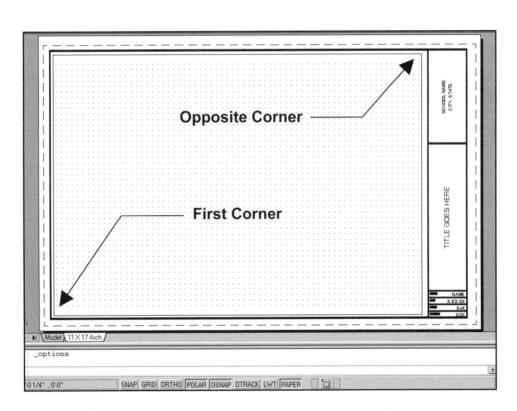

F. After successfully creating the Viewport, you should now be able to see through to Model. (Note: Turn grids OFF in paper space and ON in model space.)

G. File / Save as: **My Feet-Inches Setup.**

EXERCISE 27E
ADJUSTING THE SCALE INSIDE THE VIEWPORT

1. Open **My Feet-Inches Setup**.

2. Move to the **11 X 17 Arch** tab.

3. Double click inside of the Viewport *(It is very important that you be inside the Viewport. You want to scale the Model geometry not the Layout)*

4. Select **View / Zoom / All** (This puts your drawing limits in the correct location)

5. **Adjust the scale** of the Model geometry as follows:

   a. Select 1/4" = 1' from the **"Viewports"** toolbar.

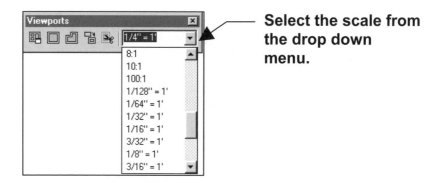

Select the scale from the drop down menu.

The Model geometry should have changed. You will not notice too much of a change other than your grid spacing should appear smaller.
Now your "11 X 17 Arch" layout is ready to accept a very large drawing.
You will understand better after you complete Exercise 27G.

6. **Lock the Viewport scale**
   Refer to page 26-11

EXERCISE 27F
PLOTTING FROM PAPER SPACE

The following instructions will guide you through the final steps for setting up the master file for plotting. These settings will stay with **My Feet-Inches Setup** and you will be able to use it over and over again. (Refer to Page 26-15 for more detailed explanations.)

A. Open **My Feet-Inches Setup** if it isn't already open.

B. Select the **11x17 Arch** layout tab.

You should be looking at your Border and Title Block now.

C. Select **File / Plot** or place the cursor on the **11x17 Arch** tab and press the right mouse button. Select **Plot** from the short cut menu.

The Plot dialog box shown below should appear.
Select the "More Options" button if your dialog box does not appear the same as shown below

D. Select the **11 X 17 Arch** Page Setup

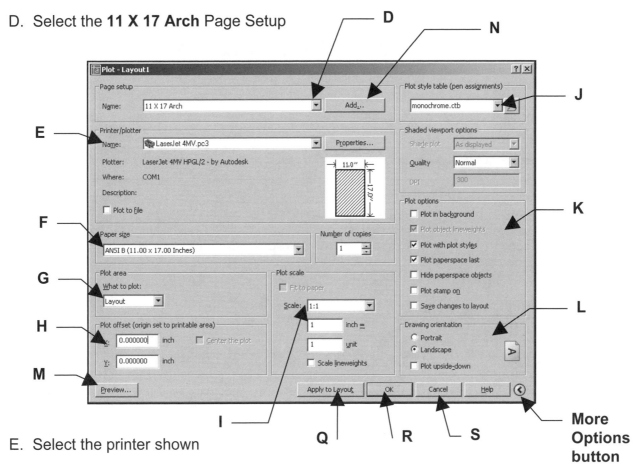

E. Select the printer shown

F. Select the Paper Size you wish to plot on.

G. Select the Plot Area

H. Plot offset should be 0 for both X and Y.

I. Select scale 1 : 1 (If you selected your printer for item "E", select "Fit to Paper")

J. Select the Plot Style Table "Monochrome.ctb"

K. Select the Plot options shown

L. Select Drawing Orientation: Landscape

M. Select **Preview** button.

    1. If the drawing is centered on the sheet, press the Esc key and continue on to **N**.

    2. If the drawing does not look correct, press the Esc key and check all your settings, then preview again.

N. Select the **ADD** button.

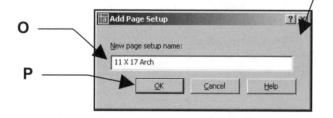

This step is actually unnecessary since we gave the page setup and layout tab the same name.
But I left it here anyway for a reference.
If you had changed any of the settings you would have to Add a new name.

O. Type the new page setup name: 11 X 17 Arch

P. Select OK button.

The settings are now saved for future use.

Q. Select the "Apply to Layout" button.

R. If your computer **is** connected to the plotter / printer selected, select the **OK** button to plot, then proceed to **T**.

S. If your computer is <u>not</u> connected to the plotter / printer selected, select the Cancel button to close the Plot dialog box and proceed to **T**. Note: Selecting Cancel <u>will cancel</u> your selected setting if you did not save the page setup as specified in **N**.

T. Save the drawing
    1. Select File / Save As
    2. Save as: **My Feet-Inches Setup**

*You have now completed the **Page Setup** for the **My Feet-Inches Setup**.   Now you are ready to use this master file to create many drawings in the future. <u>In fact, you have one on the very next page.</u>*

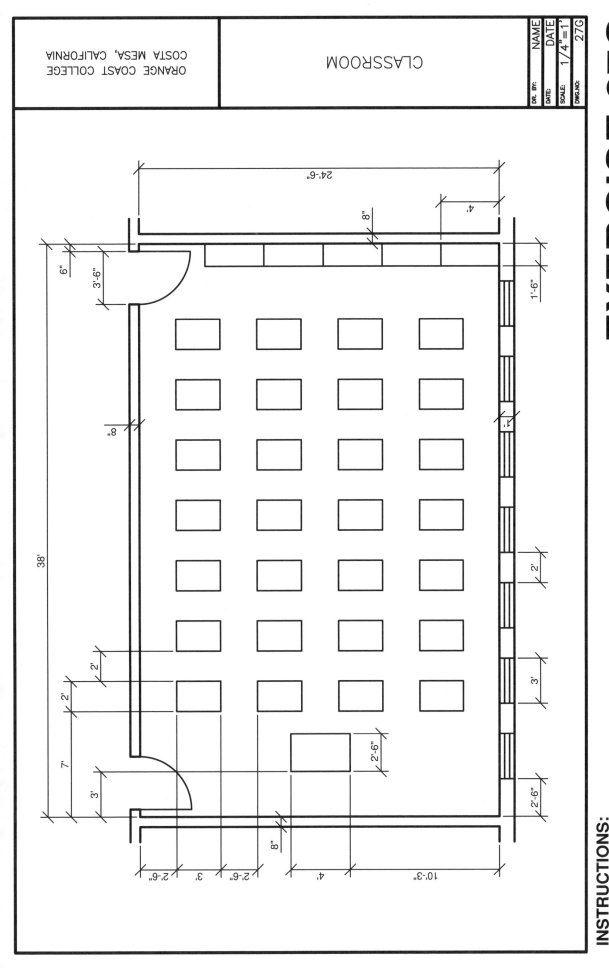

ORANGE COAST COLLEGE
COSTA MESA, CALIFORNIA

CLASSROOM

DR. BY:
NAME
DATE:
DATE
SCALE:
1/4" = 1'
DWG.NO:
27G

EXERCISE 27G

Note: If you move the drawing and the dimensions do not move with it, type **Dimregen** <enter>

INSTRUCTIONS:
1. Open **My Feet-Inches Setup.**
2. Move to the **11 X 17 Arch** tab
3. Draw the classroom, shown above, in the viewport (model space) .
4. Dimension (dimension in Paper space) (Refer to 27-10)
   Use Dimension Style = **Arch Dim**
5. Save as **EX-27G** and plot using **11 X 17 Arch** Page Setup

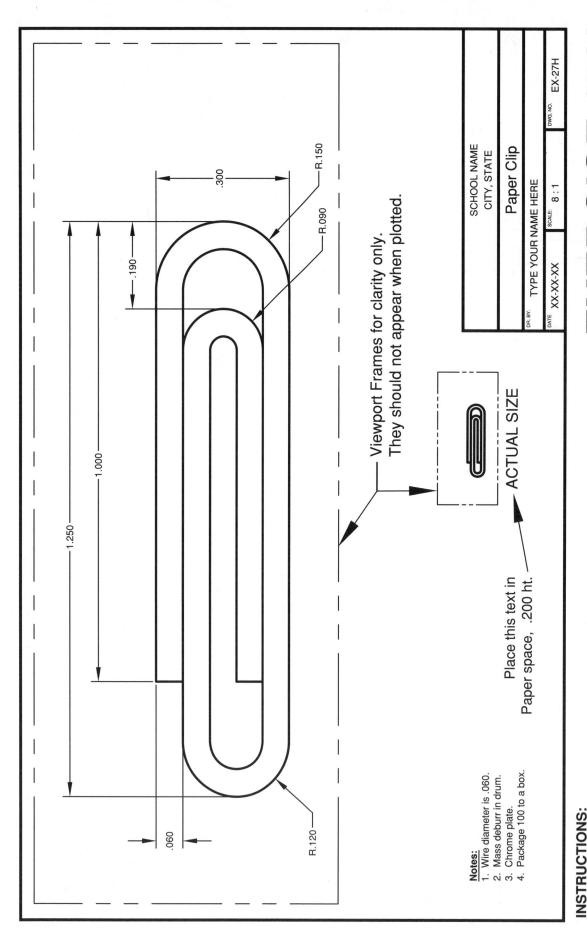

Notes:
1. Wire diameter is .060.
2. Mass deburr in drum.
3. Chrome plate.
4. Package 100 to a box.

Viewport Frames for clarity only.
They should not appear when plotted.

Place this text in
Paper space, .200 ht.

ACTUAL SIZE

SCHOOL NAME
CITY, STATE

Paper Clip

DR. BY: TYPE YOUR NAME HERE

SCALE: 8 : 1

DATE XX-XX-XX   DWG. NO. EX-27H

EXERCISE 27H

INSTRUCTIONS:

Now let's try drawing that Paper Clip that needs to be larger.

1. Open MY DECIMAL SETUP
2. Unlock the viewport.
3. Adjust the scale of the viewport to 8 : 1 and Lock again.
4. Draw the paperclip in the viewport (Model Space).
5. Dimension and Notes (1/8 ht.) in Paper space. Use Dim. style: Class style)
6. Cut a second small viewport next to the title block for the "Actual Size" dwg.
7. Double click inside the little viewport and Zoom /All.
8. Adjust the scale to 1 : 1 and lock.
   Now see why Paper space is so useful?
9. Save as: EX-27H
10. Plot using "11x17 Mono" Page Setup.

LEARNING OBJECTIVES

After completing this lesson, you will be able to:

1. Understand what are Blocks.
2. Create a Block.
3. Insert a Block into your drawing.
4. Understand the rules governing color and linetype.
5. Re-define and Purge a block.

LESSON 28

BLOCKS

A **BLOCK** is a group of objects that have been converted into ONE object. A Symbol, such as a transistor, bathroom fixture, window, screw or tree, is a typical application for the block command. First a BLOCK must be created. Then it can be INSERTED into the drawing. An inserted Block uses less file space than a set of objects copied.

CREATING A BLOCK

1. First draw the objects that will be converted into a Block.

   For this example a circle and 2 lines are drawn.

2. Select the **BMAKE** command using one of the following:

   TYPE = B
   PULLDOWN = DRAW / BLOCK / MAKE
   TOOLBAR =DRAW

   (The Block Definition dialog box will appear.)

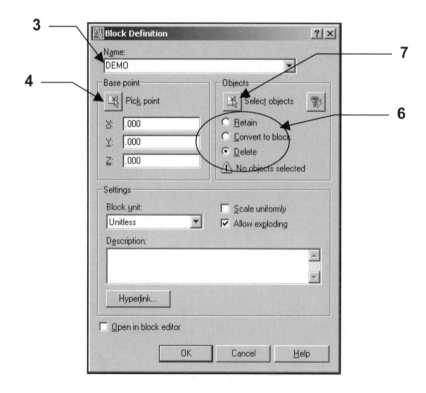

3. Enter the New Block name in the **Name** box.

4. Select the **Pick Point** button. (Or you may type the X, Y and Z coordinates.)
   The Block Definition box will disappear and you will return temporarily to the drawing.

5. Select the location where you would like the insertion point for the Block.
   Later when you insert this block, the block will appear on the screen attached to the cursor at this insertion point. Usually this point is the CENTER, MIDPOINT or ENDPOINT of an object.

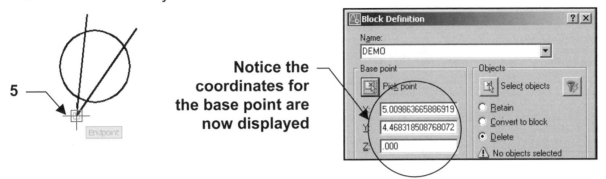

Notice the coordinates for the base point are now displayed

6. Select an option.

   It is important that you select one and understand the options below.

   Retain
   If this option is selected, the original objects will stay visible on the screen after the block has been created.

   Convert to block
   If this option is selected, the original objects will disappear after the block has been created, but will immediately reappear as a block. It happens so fast you won't even notice the original objects disappeared.

   Delete
   If this option is selected, the original objects will disappear from the screen after the block has been created.

7. Select the **Select Objects** button.

   The Block Definition box will disappear and you will return temporarily to the drawing.

8. Select the objects you want in the block, then press <enter>.

The Block Definition box will reappear and the objects you selected should be illustrated in the Preview Icon area.

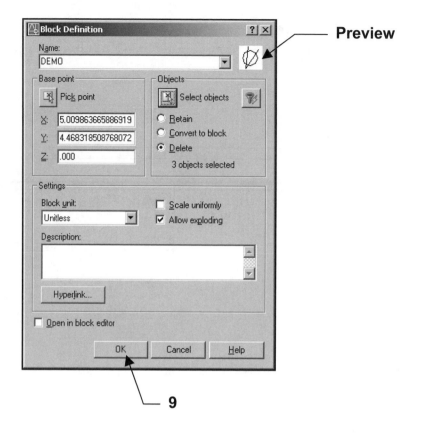

Preview

9

9. Select the **OK** button.
    The new block is now stored in the drawing's block definition table.

10. To verify the creation of this Block, select **Insert / Block**, select the Name (▼).
    A list of all the blocks, in this drawing, will appear.
    . (Refer to page 28-6)

Note: Refer to the Advanced workbook for detailed description of the option, "Dynamic Blocks".

ADDITIONAL DEFINITIONS OF OPTIONS

Block Units
You may define the units of measurement for the block. This option is used with the "Design Center" to drag and drop with Autoscaling. The Design Center is an advanced option and is not discussed in this book.

Scale Uniformly
Specifies whether or not the block is prevented from being scaled non-uniformly during insertion.

Allow Exploding
Specifies whether or not the block can be exploded after insertion.

Description
You may enter a text description of the block.

Hyperlink
Opens the **insert Hyperlink dialog box** which you can use to associate a hyperlink with the block.

HOW LAYERS EFFECT BLOCKS

If a block is created on Layer 0:

1. When the block is inserted, it will take on the properties of the current layer.
2. The inserted block will reside on the layer that was current at the time of insertion.
3. If you Freeze or turn Off the layer the block was inserted onto, the block will disappear.
4. If the Block is **Exploded**, the objects included in the block will revert to their original properties of layer 0.

If a block is created on Specific layers:

1. When the block is inserted, it will retain its own properties. It **will not** take on the properties of the current layer.
2. The inserted block **will reside** on the current layer at the time of insertion.
3. If you **freeze** the layer that was current at the time of insertion the block <u>will</u> disappear.
4. If you turn **off** the layer that was current at the time of insertion the block <u>will not</u> disappear.
5. If you **freeze** or turn **off** the blocks original layers the block <u>will</u> disappear.
6. If the Block is **Exploded**, the objects included in the block will go back to their original layer.

INSERTING BLOCKS

A **BLOCK** can be inserted at any location on the drawing. When inserting a Block you can **SCALE** or **ROTATE** it.

1. Select the INSERT command using one of the following:

    TYPE = DDINSERT
    PULLDOWN = INSERT / BLOCK
    TOOLBAR =DRAW

    The INSERT dialog box will appear.

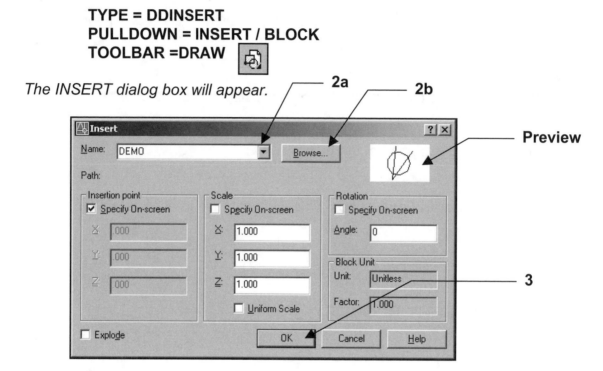

2. Select the **BLOCK** name.
    a. If the block is already in the drawing that is open on the screen, you may select the block from the drop down list shown above
    b. If you want to insert an entire drawing, select the Browse button to locate the drawing file.

3. Select the **OK** button.

    This returns you to the drawing and the selected block should be attached to the cursor.

4. Select the insertion location for the block by pressing the left mouse button or typing coordinates.

    Command: _insert
    Specify insertion point or **[Basepoint/Scale/X/Y/Z/Rotate]:**

NOTE: If you want to change the **basepoint, scale** or **rotate** the block before you actually place the block, press the right hand mouse button and you may select an option from the menu or select an option from the command line menu shown above.

You may also "preset" the insertion point, scale or rotation. This is discussed on page 28-7.

PRESETTING THE <u>INSERTION POINT</u>, <u>SCALE</u> or <u>ROTATION</u>

You may preset the **Insertion point, Scale or Rotation** in the <u>INSERT</u> box instead of at the command line.

1. Remove the check mark from any of the **"Specify On-screen"** boxes.
2. Fill in the appropriate information describe below:

Insertion point
Type the X and Y coordinates <u>from the Origin</u>. The Z is for 3D only.
The example below indicates the block's insertion location will be 5 inches in the X direction and 3 inches in the Y direction, from the Origin.

Scale
You may scale the block proportionately by typing the scale factor in the X box and then check the <u>Uniform Scale box</u>.
If the block will be scaled non-proportionately, type the different scale factors in both X and Y boxes.
The example below indicates that the block will be scale proportionate at a factor of 2.

Rotation
Type the desired rotation angle relative to its current rotation angle.
The example below indicates the block will be rotated 45 degrees from its originally created angle orientation.

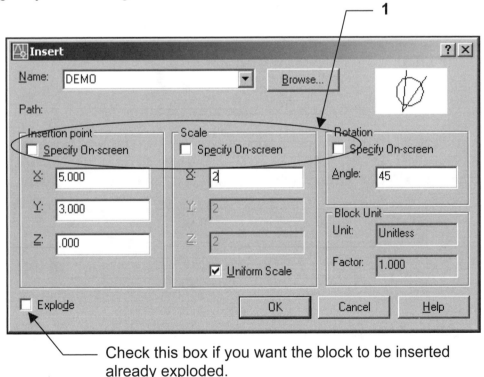

Check this box if you want the block to be inserted already exploded.

Blocks have many uses and will be discussed further in the "Exercise Workbook for Advanced AutoCAD 2007", along with "DesignCenter" and "Tool Palettes".

RE-DEFINING A BLOCK

How to change the design or insertion point of a block previously inserted.

1. Double click on the Block that you wish to change.

   *The "**Edit Block Definition**" dialog box will appear.*

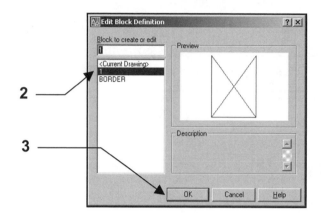

2. Select the name of the Block that you wish to change.

3. Select the **OK** button.

4. The "**Do you want to....**" dialog box appears. **Select <u>NO</u> for now**.

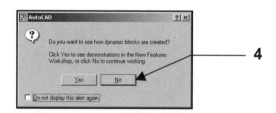

5. The Block that you selected should appear large on the screen. <u>Make the changes</u>.

6. Select the **Close Block Editor** button. (Located at the top of the screen on the tool bar.)

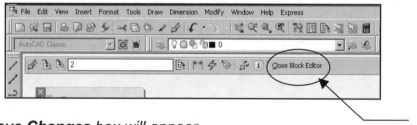

   *The **Save Changes** box will appear.*

7. Select **Yes**. You will be returned to the drawing and **ALL** of the blocks with the same name will be updated with the changes that you made.

PURGE UNWANTED BLOCKS

How to delete unwanted blocks from the current drawing.

1. Select **File / Drawing Utilities / Purge**
2. Select the **+** sign beside **Blocks**
3. Select the block that you wish to delete.
4. Select **Purge** button.
5. Select **Close** button.

> Note: You can't purge a block that is in use within the drawing.

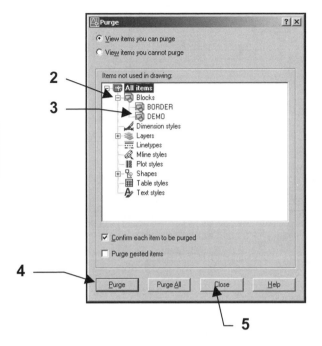

WHERE ARE BLOCKS SAVED?

When you create a Block it is saved **within the drawing you created it in**.
(If you open another drawing you will not find that block.)

In the Advanced workbook you will learn how to use the DesignCenter to drag a block from one drawing to another drawing. You will also learn about Dynamic Blocks.

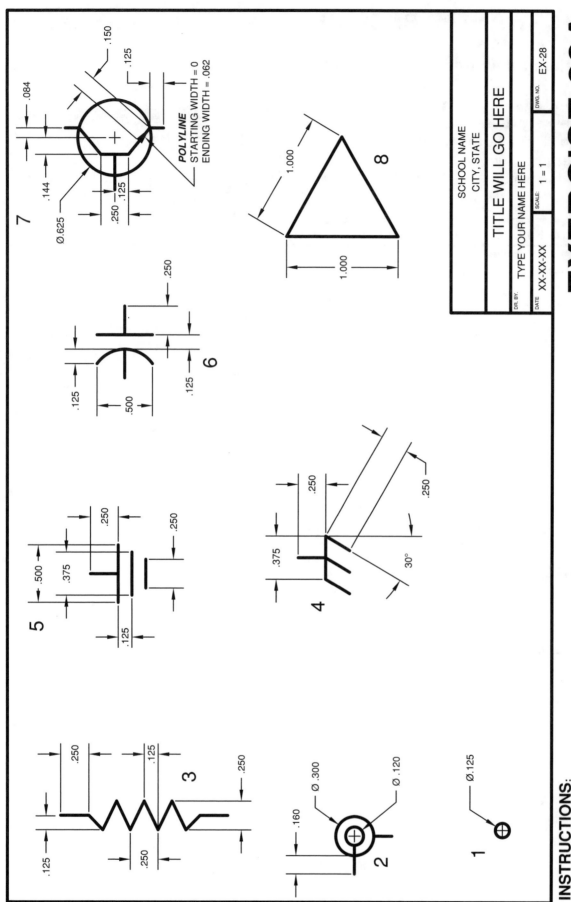

7

.150
.125
.084
.144
.250 .125
Ø.625

POLYLINE
STARTING WIDTH = 0
ENDING WIDTH = .062

8

1.000
1.000

6

.250
.125
.500
.125

5

.250
.500
.375
.125
.250

4

.250
.375
.250
30°

3

.250
.125
.125
.250
.250

2

Ø .300
Ø .120
.160

1

Ø.125

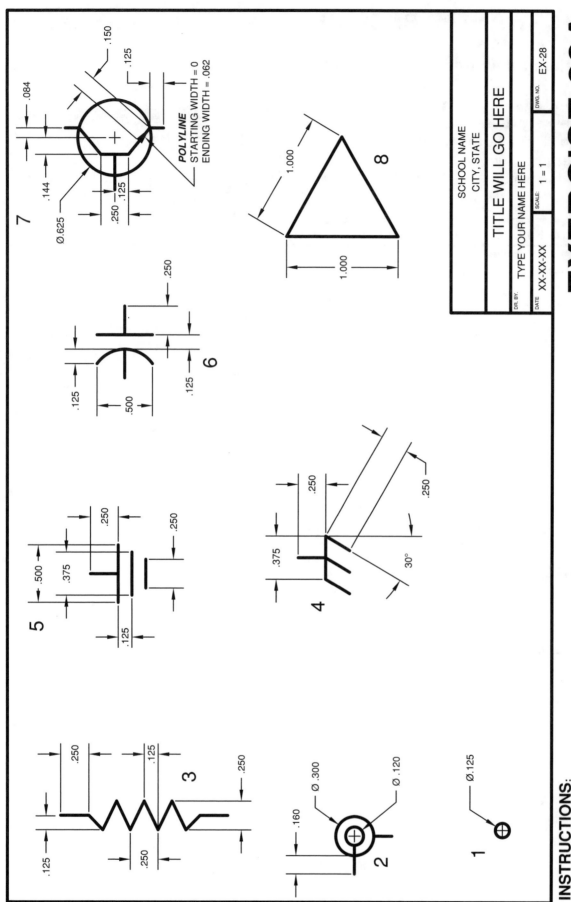

SCHOOL NAME
CITY, STATE

TITLE WILL GO HERE

TYPE YOUR NAME HERE

DR. BY.

DATE XX-XX-XX

SCALE: 1 = 1

DWG. NO. EX-28

EXERCISE 28A
STEP 1

INSTRUCTIONS:

1. Open **My Decimal Setup.**
2. Draw the Objects above inside the viewport (Model space) Use Layer = Object.
3. Now change each one into a Block. (They should disappear as they are made)
4. Name them 1, 2, 3, etc.
5. Save as **EX-28A**
6. Now follow the instructions for Step 2 on the next page. (Do not start a new drawing)

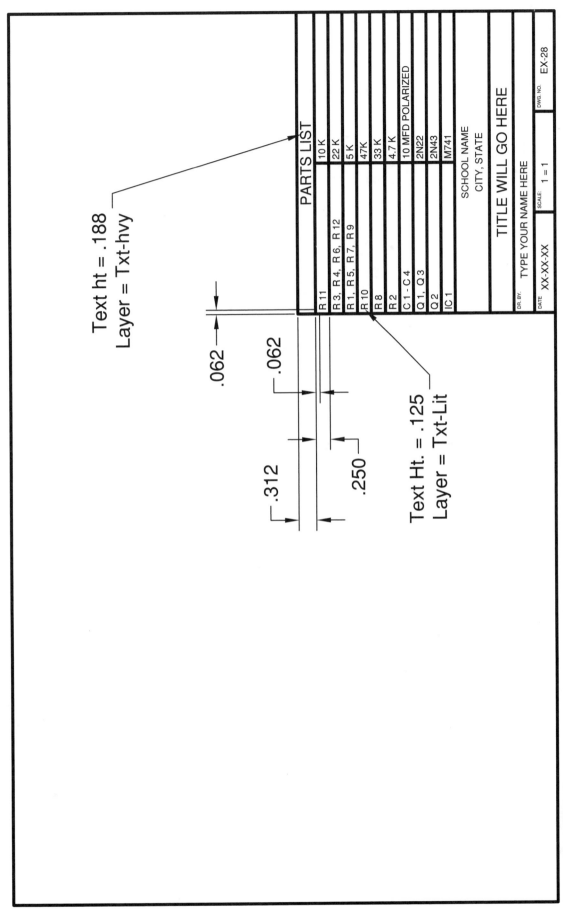

Text ht = .188
Layer = Txt-hvy

PARTS LIST

R 11	10 K
R 3, R 4, R 6, R 12	22 K
R 1, R 5, R 7, R 9	5 K
R 10	47K
R 8	33 K
R 2	4.7 K
C 1 - C 4	10 MFD POLARIZED
Q 1, Q 3	2N22
Q 2	2N43
IC 1	M741

.062

.062

.312

.250

Text Ht. = .125
Layer = Txt-Lit

	SCHOOL NAME	
	CITY, STATE	
	TITLE WILL GO HERE	
DR. BY:	TYPE YOUR NAME HERE	DWG. NO. EX-28
DATE XX-XX-XX	SCALE: 1 = 1	

EXERCISE 28A
STEP 2

INSTRUCTIONS:

1. Open **EX-28A** (if it isn't already open).
2. Create the Parts List above the title block in paper space 17X11(1 to 1) tab
3. Use Layer **Border** for the lines and Layer **Text-Lit** for the text.
4. Save as **EX-28A again.**
5. Now follow the instructions for Step 3 on the next page. (Do not start a new drawing.)

28-11

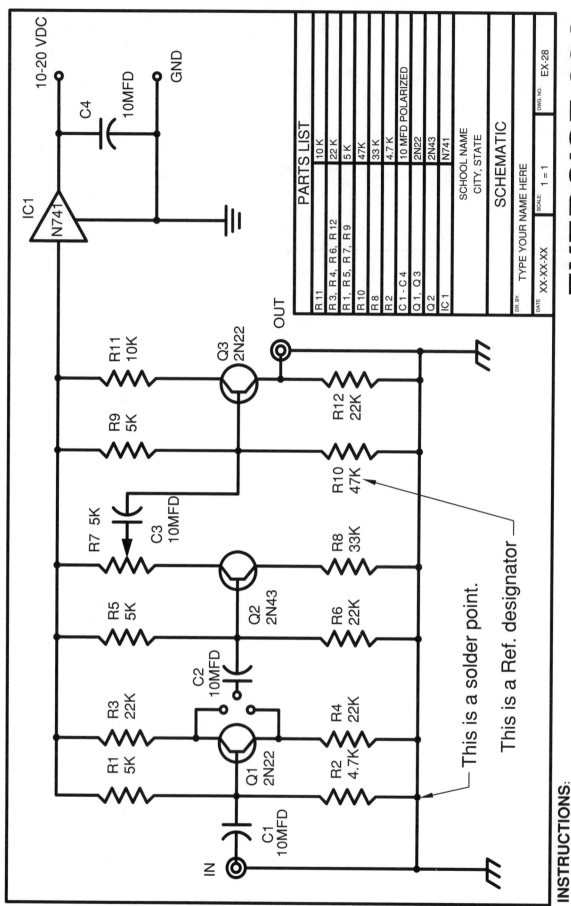

EXERCISE 28A
STEP 3

PARTS LIST

R 11	10 K	
R 3, R 4, R 6, R 12	22 K	
R 1, R 5, R 7, R 9	5 K	
R 10	47 K	
R 8	33 K	
R 2	4.7 K	
C 1 - C 4	10 MFD POLARIZED	
Q 1, Q 3	2N22	
Q 2	2N43	
IC 1	N741	

SCHOOL NAME
CITY, STATE

SCHEMATIC

| DR. BY. | TYPE YOUR NAME HERE | SCALE: 1 = 1 | DWG. NO. EX-28 |
| DATE | XX-XX-XX | | |

This is a solder point.

This is a Ref. designator

INSTRUCTIONS:

1. Open **EX-28A** (If it isn't already open).
2. Draw the schematic above in modelspace. Use Layer Object.
3. Use Donuts to make the Solder points, ID=0   OD=.125.
4. Add Reference Designators on Layer Txt-Lit. Ht =.125
5. Save as **EX-28A** again and Plot using Page Setup **"11x17 Mono"**.

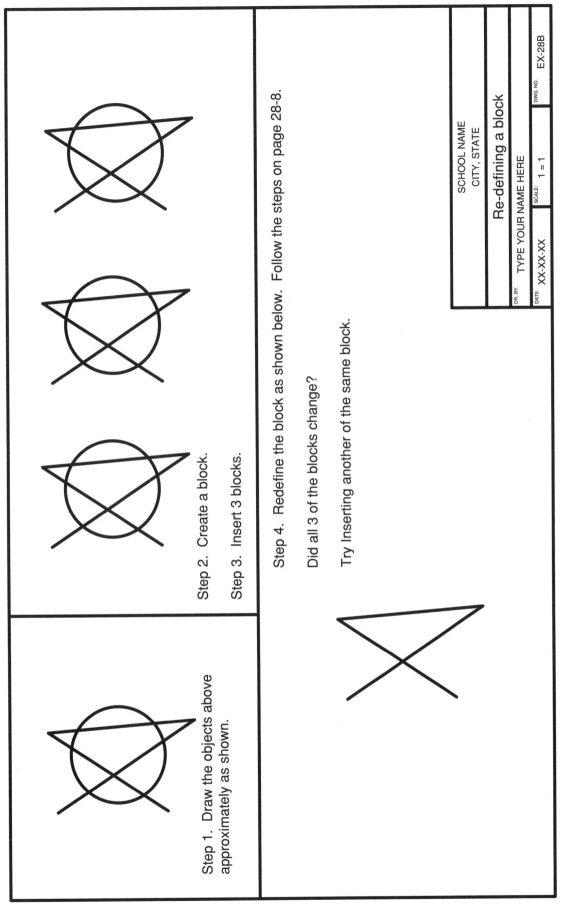

Step 1. Draw the objects above approximately as shown.

Step 2. Create a block.

Step 3. Insert 3 blocks.

Step 4. Redefine the block as shown below. Follow the steps on page 28-8.

Did all 3 of the blocks change?

Try Inserting another of the same block.

SCHOOL NAME
CITY, STATE

Re-defining a block

DR. BY. TYPE YOUR NAME HERE
SCALE: 1 = 1
DWG. NO. EX-28B
DATE XX-XX-XX

EXERCISE 28B

INSTRUCTIONS:
1. Open **My Decimal Setup.**
2. Follow the steps shown above.
3. Save as **EX-28B**
Do not plot. This is just an example of how the "Re-define" feature works.

NOTES:

LEARNING OBJECTIVES

After completing this lesson, you will be able to:

1. Create a multiview layout for plotting.
2. PAN a drawing image within a viewport.

LESSON 29

PAN

After you adjust the scale of a viewport sometimes the drawing, within the viewport frame, is not placed as you would like it. If you use Zoom / All, the scale will not be correct and you would need to re-adjust the scale. (Refer to the example on page 29-3)

This is where the **PAN** command comes in handy. The PAN command will allow you to move the drawing around, within the viewport, without affecting the adjusted scale.

Note: Do not use the MOVE command. You do not want to actually move the original drawing. You only want to slide the viewport image, of the original drawing, around within the viewport.

This will make more sense after you have completed Exercise 29, step 3.

How to use the PAN command.

1. Unlock the viewport.

2. Activate the viewport. (Double click inside viewport)

3. Select the **PAN** command using one of the following:

> **TYPE = P**
> **PULLDOWN = VIEW / PAN / REALTIME**
> **TOOLBAR = STANDARD**

4. Place the cursor inside the viewport and hold the left mouse button down while moving the cursor. (Click and drag) When the drawing is in the desired location release the mouse button.

4. Lock the viewport.
    Now you can use the Zoom commands and it will not affect the adjusted scale.

Note: You may set the "wheel" on your wheel mouse to Pan when you press and hold.
       Refer to page Intro-10 for the Mbuttonpan settings.

Example of before and after panning.

BEFORE PANNING

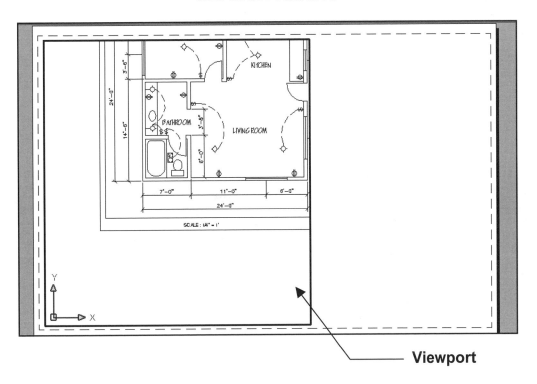

Viewport

AFTER PANNING

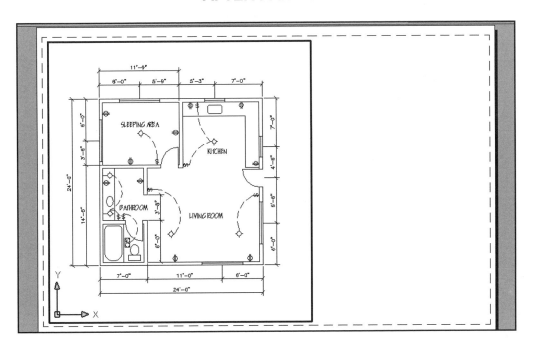

EXERCISE 29

This exercise will give you more practice using Blocks, working with a scaled drawing, using multiple viewports and using the PAN command.

STEP 1
Create Blocks

A. Open your drawing **My Feet-Inches Setup.**

B. **Create the Blocks**
Select the "**Model**" tab and draw the objects shown on page 29-5 in **Model Space**. (Use the layers indicated.)

Note: your blocks will appear smaller than the example. I enlarged them for clarity.

Change each one into a **BLOCK**. They should disappear as you create the Blocks. (Refer to Lesson 28 for assistance.)

C. Save the drawing as **EX-29.**

Do not start a new drawing. Continue on to Step 2.

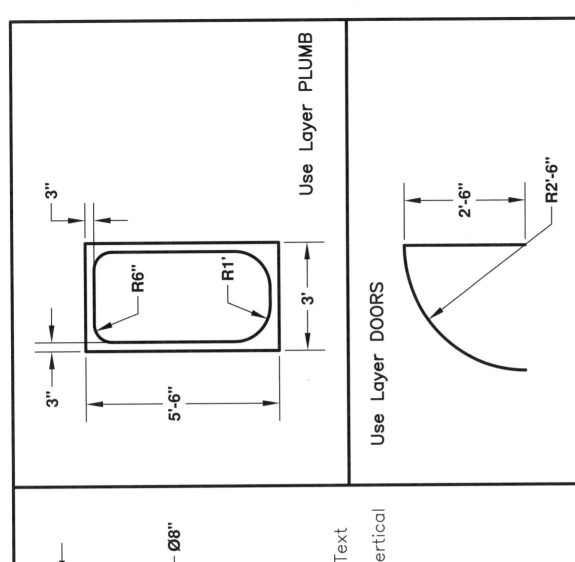

Use Layer PLUMB

Use Layer DOORS

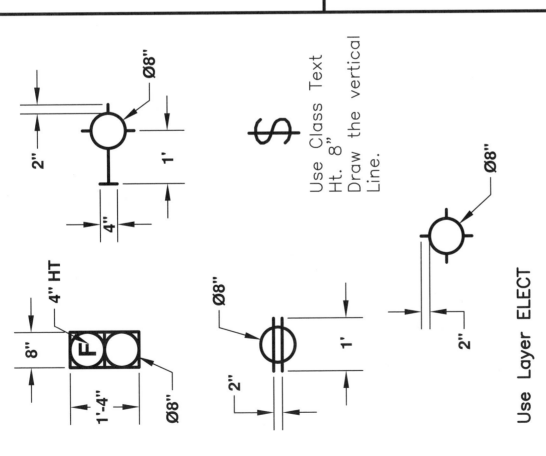

Use Class Text
Ht. 8"
Draw the vertical
Line.

Use Layer ELECT

EXERCISE 29
STEP 1

EXERCISE 29

STEP 2
Draw the Cabin

A. Select the **"11 X 17 Arch"** tab. (You should see your title block.)
(Next double click in the viewport to activate the Model space.)

B. Draw the **CABIN** shown on page 29-7 in <u>Model Space</u>.

Insert the Blocks you made in Step 1.

Design your own Kitchen. Use appropriate layers.

Walls = 6" thick   **Space behind doors** = 2"
Window sizes: In the Kitchen = 3 ft. In the Sleeping and Living rooms = 6 ft.

C. How to draw the dashed Electrical Lines.

   1. Use Layer **Wiring**.
   2. Draw the lines using **Arc** (Start, End, Direction).
   3. Change the **Ltscale** to **.5** using one of the options below:

   <u>Globally:</u>
   -Type at the command line: ***Ltscale <enter>.***
   -Enter new linetype scale factor <1.0000>:***type .5 <enter>.***

   OR

   <u>Individually</u>: Select the arcs individually and use the Properties Palette.

   (Refer to page 27-11 for information on Linetype scales.)

IMPORTANT:
D. Dimension the drawing as shown.   Use dim style: Arch Dim.

   Dimension in <u>Model Space this time</u>. (Note: The DSF is 48. Refer to pg 27-7.)
   (Try to place the dimensions as shown.)

E. Save as **EX-29**.

<u>**Do not start a new drawing. Continue on to Step 3.**</u>

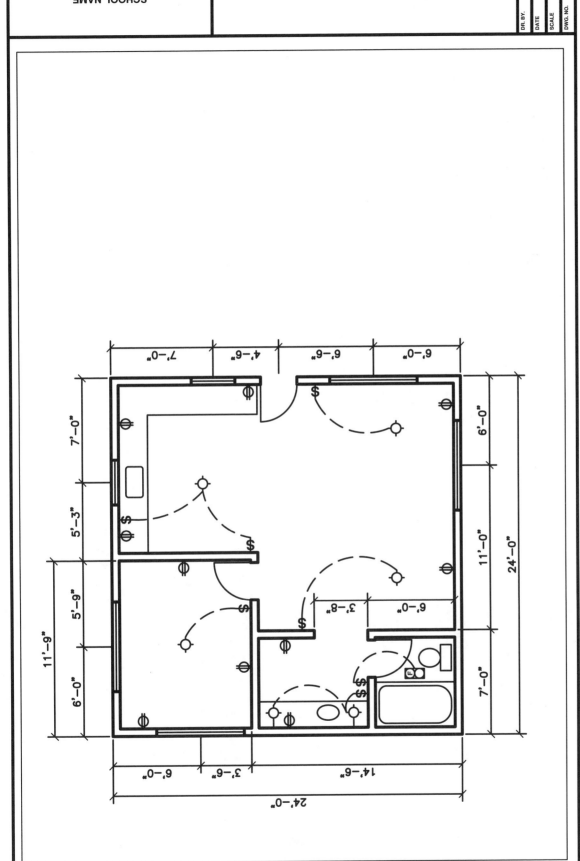

SIERRA CABIN FLOOR PLAN

SCHOOL NAME
CITY, STATE

NAME
DR. BY.
DATE   X-XX-XX
SCALE   1/4"=1'
DWG. NO.   29

EXERCISE 29

STEP 3
Creating Multiple Viewports

A. Unlock the viewport and erase it. (Just click on the VP frame & select Erase.) The drawing will disappear because the viewport is gone and you can't see through to Model Space. But your border should still be there.

B. Cut 3 new viewports approximately as shown. Sizes shown below.

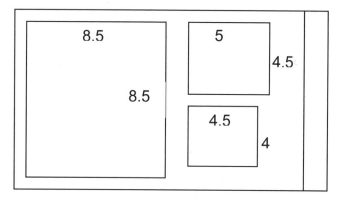

You should see your cabin drawing in each of the viewports.

D. Adjust the scale of each viewport. (Refer to the scales shown under each viewport on page 29-9.)

E. Use the PAN command to move the cabin within each viewport to the approximate location shown.

F. Lock the viewports.

G. Add the text in Paper Space.
Scales = Ht. = 1/8"   **Room Names** = Ht = 3/16"

H. Before you plot this drawing, go to **FORMAT / LAYERS** and change the Viewport layer to "Plotable". In this case, I think the viewport frame should be visible.

I. Save the drawing as **EX-29** and plot using page setup **11 X 17 Arch.**

Think about what you just accomplished.
You drew one drawing but created a multiview drawing by using paper space and viewports. And all views are scaled to a specific scale.

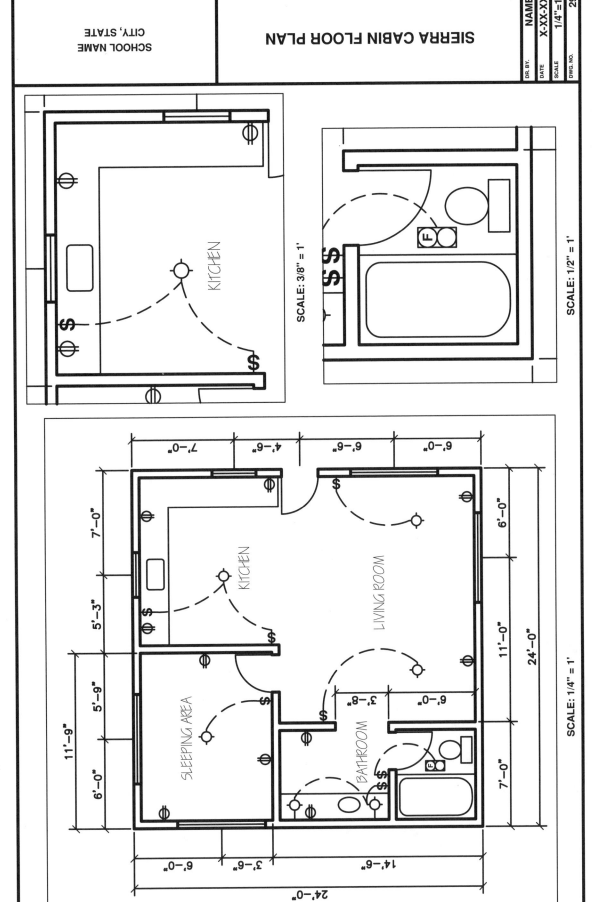

SIERRA CABIN FLOOR PLAN
SCHOOL NAME
CITY, STATE

DR. BY.		NAME
DATE		X-XX-XX
SCALE		1/4"=1'
DWG. NO.		29

KITCHEN

SCALE: 3/8" = 1'

SCALE: 1/2" = 1'

KITCHEN

LIVING ROOM

SLEEPING AREA

BATHROOM

7'-0"

5'-3"

5'-9"

11'-9"

6'-0"

7'-0"

4'-6"

6'-6"

6'-0"

6'-0"

11'-0"

24'-0"

3'-8"

6'-0"

7'-0"

6'-0"

3'-6"

14'-6"

24'-0"

SCALE: 1/4" = 1'

EXERCISE 29
STEP 3

NOTES:

LEARNING OBJECTIVES

In this lesson we will review and practice:

1. "Polyline / Spline" commands. (Lesson 23)

LESSON 30

EXERCISE 30

INSTRUCTIONS:

A. Open drawing **My Decimal Setup**
   Select the 11 X 17(1 to 1) tab. (You should see your title block)
   Double click in the viewport area to get into model space.

B. **STEP 1**.
   Draw the details shown on page 30-3. <u>Do not dimension</u>.
   Draw the break line as follows:
   a. Draw the break line using **POLYLINE**.
   b. Change it to a curved line using:
      MODIFY / OBJECT / POLYLINE
         Select the Polyline
         Select **S**pline

   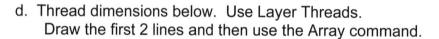

   c. Draw the Hatch
      Use Hatch pattern ANSI 31 and Layer = Hatch

   d. Thread dimensions below. Use Layer Threads.
      Draw the first 2 lines and then use the Array command.

   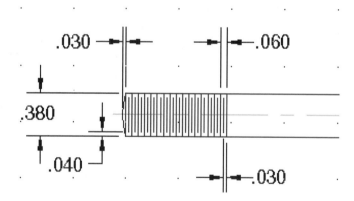

C. **STEP 2.**
   Return to paperspace.
   Draw the **PARTS LIST** (in paperspace) shown on page 30-4, STEP 2.
   Make the changes to the title block.

D. **STEP 3.**
   Move the parts into the proper locations to form the assembly shown.
   The distance between the jaws is 1".
   Draw the Balloons:
   a. Use Leader
   b. Use a .50 dia circle.
   c. Text ht. = .187

E. Save as **EX-30** and **PLOT** using Page setup "**11 x 17 Mono**".

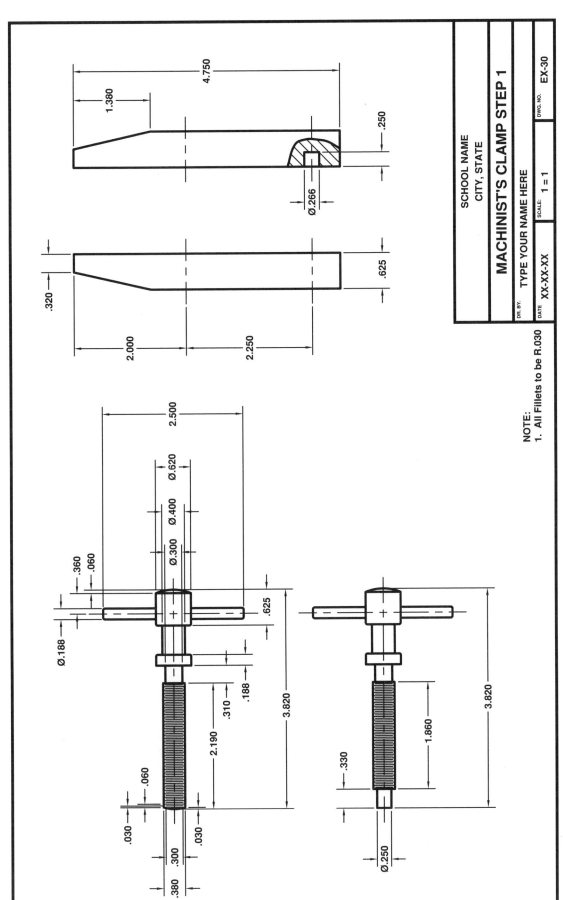

SCHOOL NAME
CITY, STATE

MACHINIST'S CLAMP STEP 1

DR. BY. TYPE YOUR NAME HERE

DATE XX-XX-XX   SCALE: 1 = 1   DWG. NO. EX-30

NOTE:
1. All Fillets to be R.030

EXERCISE 30
STEP 1

Follow Step 1 instructions on page 30-2.

30-3

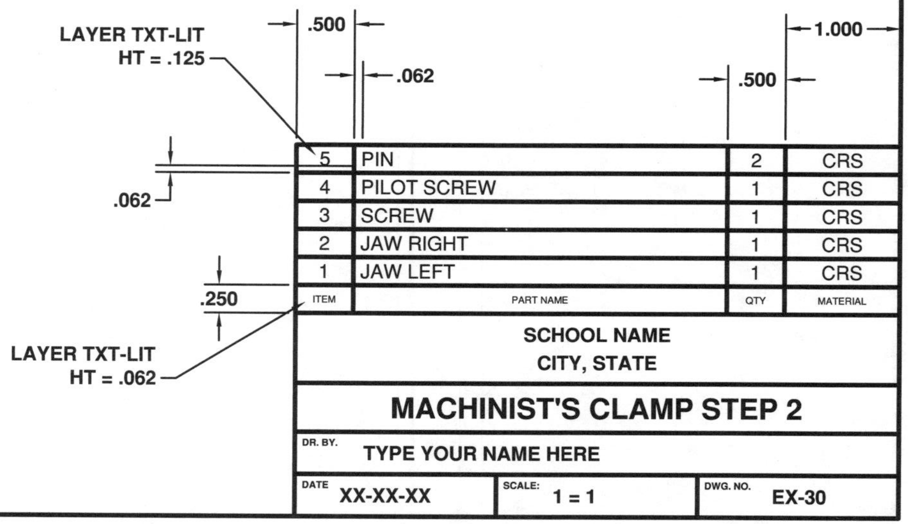

EXERCISE 30
STEP 2

Follow Step 2 instructions on page 30-2.

ITEM	PART NAME	QTY	MATERIAL
5	PIN	2	CRS
4	PILOT SCREW	1	CRS
3	SCREW	1	CRS
2	JAW RIGHT	1	CRS
1	JAW LEFT	1	CRS

SCHOOL NAME
CITY, STATE

MACHINIST'S CLAMP STEP 2

DR. BY.	TYPE YOUR NAME HERE		DWG. NO. EX-30
DATE XX-XX-XX		SCALE: 1 = 1	

1.000

.500

.062

.500

.062

.250

LAYER TXT-LIT
HT = .125

LAYER TXT-LIT
HT = .062

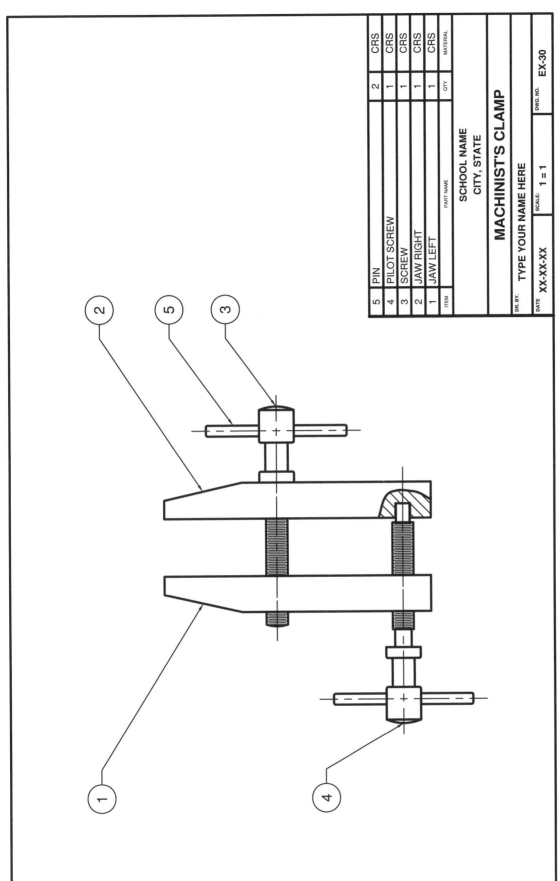

ITEM	PART NAME	QTY	MATERIAL
5	PIN	2	CRS
4	PILOT SCREW	1	CRS
3	SCREW	1	CRS
2	JAW RIGHT	1	CRS
1	JAW LEFT	1	CRS

SCHOOL NAME
CITY, STATE

MACHINIST'S CLAMP

DR. BY.	TYPE YOUR NAME HERE		DWG. NO.
DATE	XX-XX-XX	SCALE: 1 = 1	EX-30

EXERCISE 30
STEP 3

Follow Step 3 instructions on page 30-2.

NOTES:

APPENDIX A
Add a Printer / Plotter

The following are step-by-step instructions on how to configure AutoCAD for your printer or plotter. These instructions assume you are a single system user. If you are networked or need more detailed information, please refer to your AutoCAD users guide.

Note: You can configure AutoCAD for multiple printers. I suggest that you configure the plotter shown below to match the exercises in this workbook.

A. Select **File / Plotter Manager**
B. Select **"Add-a-Plotter"** Wizard

B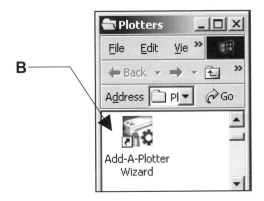

C. Select the **"Next"** button.

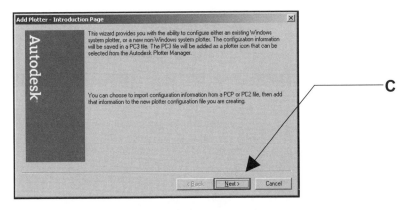 C

D. Select **"My Computer"** then **Next**.

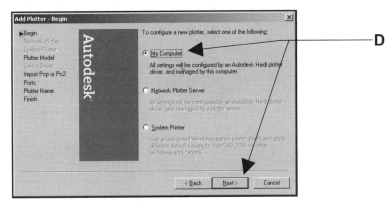 D

E. Select the **Manufacturer** and the specific **Model** desired then **Next**.

(If you have a disk with the specific driver information, put the disk in the disk drive and select "Have disk" button then follow instructions.)

NOTE: Please configure this printer on your system. This <u>will not</u> effect your computer in any negative way.

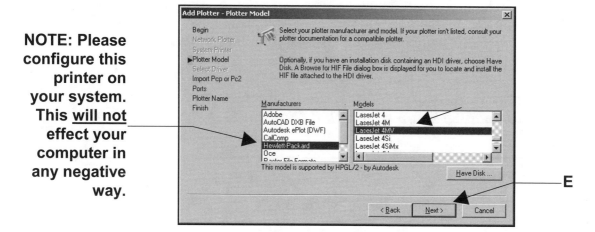

E

F. Select the **"Next"** box.

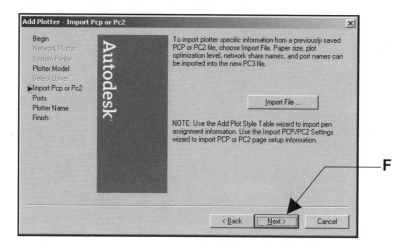

F

G. Select **"Plot to a port"**. Then select **"Next"**.

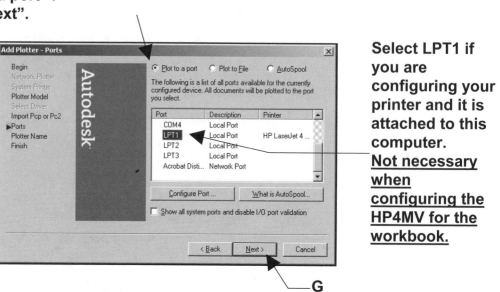

Select LPT1 if you are configuring your printer and it is attached to this computer. <u>Not necessary when configuring the HP4MV for the workbook.</u>

G

H. The Printer name, that you previously selected, should appear. Then select **"Next"**

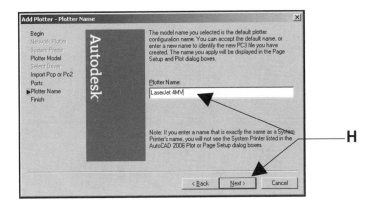

I. Select the **"Edit Plotter Configuration…"** box.

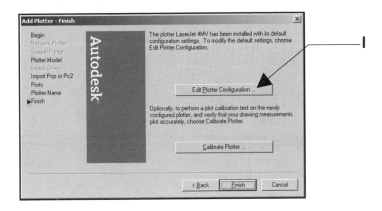

J. Select:
1. Device and Document Settings tab.
2. Media: Source and Size
3. Size: Ansi B (11 X 17 inches)
4. OK box.

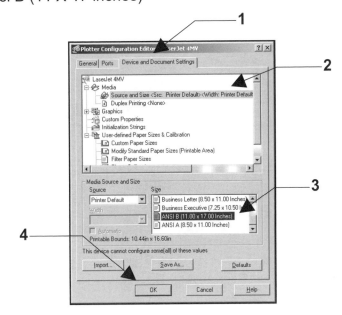

Appendix-A3

K. Select **"Finish"**.

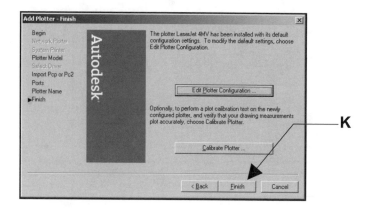

L. Now check the **File / Plotter Manager.**

Is the printer / plotter there?

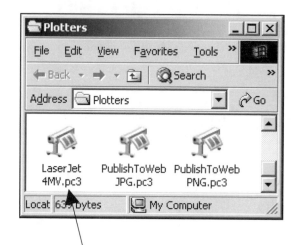

Note: It is important that this printer "HP LaserJet 4MV" is configured on your system in order to complete the exercises. <u>This will not harm you system.</u>

APPENDIX B
Dimension Style Definitions

When you select **Dimension / Style / Modify** the following dialog box will appear.

The following are descriptions for each setting within each section tab.

Lines tab

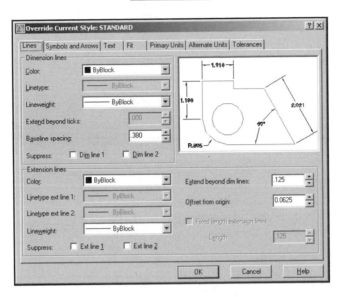

Dimension Lines
Color: Color of the dimension line.
Linetype: Sets the linetype for the dimension lines.
Lineweight: Sets the width of the dimension lines.
Extend beyond ticks: Distance to extend <u>dimension</u> line beyond the <u>extension</u> line.
Baseline spacing: Spacing between baseline dimensions.
Suppress: Suppress means disappear. You may suppress individually or both.

Extension Lines
Color: Color of the extension lines
Linetype ext line 1: Sets the linetype for the 1st extension lines.
Linetype ext line 2: Sets the linetype for the 2nd extension lines.
Lineweight: Sets the width of the extension lines.
Supress: You may suppress the 1st or 2nd extension lines individually or both.
Extend beyond dim. line: Distance to extend <u>extension</u> line beyond the <u>dimension</u> line. Example above.
Offset from origin: The distance between the object and the extension line. Example above.
Fixed length extension lines: Sets the total length of the extension lines starting from the dimension line toward the dimension origin.

Symbols and Arrows tab

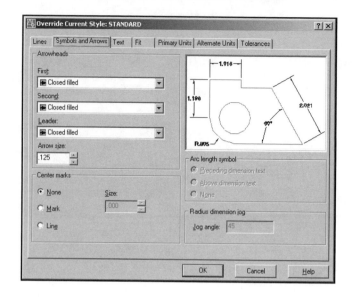

Arrowheads

First : Sets the style for the symbol inserted for the first extension point

Second: Sets the style for the symbol inserted for the second extension point and for Diameter and Radius.

Leader: Sets the style for the symbol inserted for Leaders.

Arrow Size: Sets the Size of Arrowheads.

Center Marks for Circles

None: No center marks.

Mark: criss cross in the center of the circle

Line: Mark plus lines extending beyond the circle.

Size: Specify Size.

Text tab

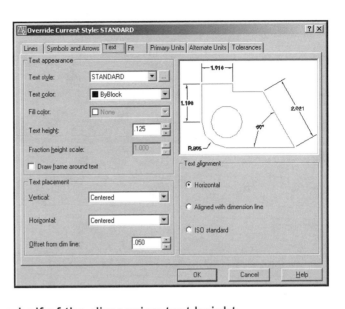

Text Appearance

Text Style: Select which text style to use for the dimension text. To create or change a style, select the [...] button.
Text Color: Sets the color of the dimension text.
Fill Color: Sets the color for the text background in dimensions.
Text height: Sets the height.
Fraction Height Scale: Sets the size of fractions relative to the dimension text height. This is a factor not and actual height. Example: A setting of .50 would be half of the dimension text height.
Draw Frame Around Text: If this box is checked, a box will be drawn around the text.

Text Placement

Vertical
Centered: Centers dimension text between extension lines.
Above: Places dimension text above the dimension line.
Outside: Places dimension text on the side of the dimension line farthest away from the object.

Horizontal
Centered: Centers the dimension text along the dimension line between the ext lines.
At Ext Line 1: Moves text near first extension line.
At Ext Line 2: Moves text near second extension line.
Over Ext Line 1: Places text over first extension line.
Over Ext Line 2: Places text over second extension Line

Offset from Dim Line: Sets the gap between the dimension text and the dimension line.

Text Alignment
Horzontal: Places dimension text horizontal.
Aligned: Aligns dimension text with the dimension line.
ISO Standard: Aligns text with the dimension line when text is inside the extension lines, but aligns it horizontally when text is outside extension lines.

FIT tab

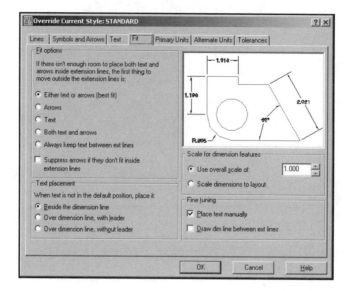

Fit Options: Controls the placement of text and arrowheads based on the space available between the extension lines.   If there is not enough room to place both text and arrows inside extension lines, you must choose what to move outside the extension lines.

Text Placement: When dimension text is moved from its dim. style location, this setting controls the placement.

Scale for Dimension Features:

Use overall scale of: Set the factor for all dimension settings. Example: if this is set to 10 and the dimension text height is set to 1/8, the dimension text height would appear 1-1/4" ht. (10 x 1/8")

Scale dimensions to layout (paper space): Calculates a scale factor based on the scaling of model space vs. paper space.

Fine Tuning

Place text manually when dimensioning: You control the placement of the text if this feature is ON. It is not automatic.

Draw dim Line between Ext Lines: Draws the dimension line inside the extension lines even if the arrows are on the outside.

Primary Units tab

Linear Dimensions

Unit Format: Sets the units format for all dimensions except Angular.

Precision: Sets the number of decimal places in the dimension.

Fraction Format: Set the format for fraction to Horizontal, Diagonal or not stacked. Only available if Unit Format (above) is set to "Fractional".

Decimal Separator: Sets the style for the decimal separator to period, comma or space. Not available if Unit Format is set to Fractional.

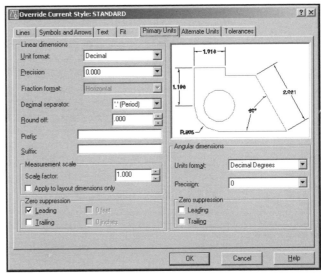

Round Off: Sets rounding limits for dimensioning, such as .000 or 1/8. Rounds **up** to the nearest 3 place decimal or nearest 1/8".

Prefix: Add text in front of the dimension text. (**Flat for** 2.00)

Suffix: Add text after the dimension text. (1'-0" **Max**)

Measurement scale:

Scale Factor: AutoCAD multiples the dimension measurement by the value entered here. Example: If you draw a 1/2 inch line. Set this feature to 2. When you dimension the line, the dimension text will display 1. (2 X 1/2) If you set this feature to .50, when you dimension the line, the dimension text will display 1/4. (.50 X 1/2)

Apply to Layout Dimensions Only: Unnecessary now that we have Trans-spatial dimensioning.

Zero Suppression

The following two only work with decimals:

Leading: Controls the display of zeros before the decimal point. 0.50 = Off   .50 = ON

Trailing: Controls the display of zeros at the end of the dimension. .500 = Off .5 = ON

The following two only work with architectural:

Feet: Controls the display of zeros for feet. 0'-6" = OFF   6" = ON

Inches: Controls the display of zeros for inches. 6'-0" = OFF   6" = ON

Angular Dimensions

Units Format: Sets the units format for Angular. Does not affect Linear.

Precision: Sets the number of decimal places past the whole degree.

Zero suppression: Same as Linear.

Note: Alternate Units and Tolerance tabs are discussed in the "Advanced" workbook.

Notes:

APPENDIX C
ASSIGN LINEWEIGHTS TO COLORS

In the 1workbook helper.dwg, lineweights have been assigned to layers. When you select the appropriate layer the lineweight is automatically drawn. For example: when you select the layer "object" the lineweight will be .031. This lineweight will be visible on the screen and will plot.

Lineweights may also be assigned to colors within the "color dependent plot style table". Some AutoCAD users prefer to assign lineweights to colors rather than layers or individual objects. As you get more familiar with Lineweights and plotting, you can determine which process you prefer.

To assign lineweights to colors, you must create a "Color dependent Plot Style Table". When this plot style table is selected, it will override the lineweights assigned within the drawing. The step by step process is described below.

1. Select **FILE / PLOT STYLE MANAGER**

2. Select **"Add-A-Plot Style Table" Wizard**

The following dialog boxes will appear.

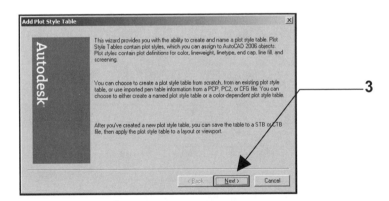

3. Select the **Next** button.

4. Select **"Start from Scratch"** then the **Next** button.

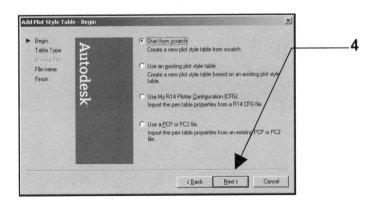

Appendix C-1

5. Select **"Color-Dependent Plot Style Table"** then the **Next** button.

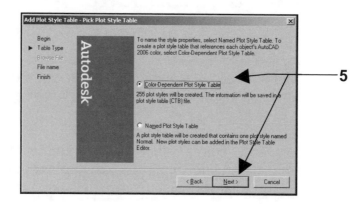

6. Type the new Plot Style Table **name** then select the **Next** button.

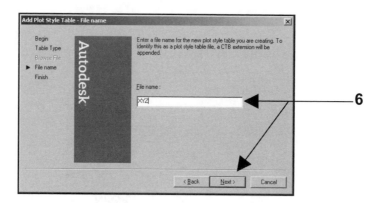

7. Select the **"Plot Style Table Editor"** button.

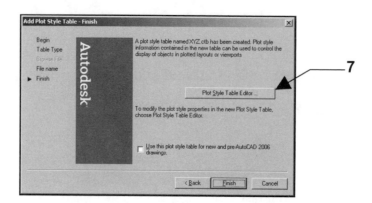

8. Make changes to the **"PROPERTIES"** then select the **Save & Close** button.

8a. Select color 1

8b. Change Lineweight for the color selected.

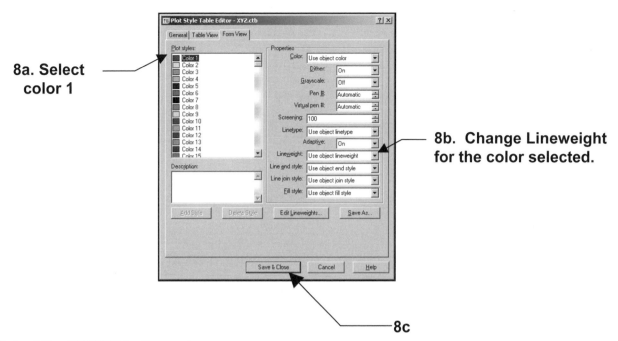

8c

9. Select the **FINISH** button.

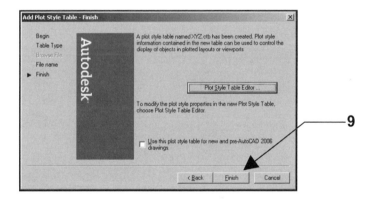

9

10. Select **File / Plot Style Manager**.

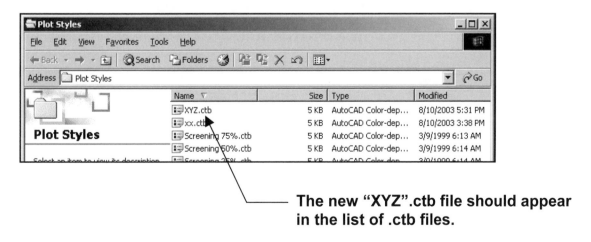

The new "XYZ".ctb file should appear in the list of .ctb files.

NOTES:

APPENDIX D
BACKGROUND PLOTTING

Background Plotting allows you to continue to work while your drawing is plotting. This is a valuable time saver because some drawings take a long time to plot. Or maybe you have multiple drawings to plot and you do not want to tie up your computer.

If you wish to view information about the status of the plot, click on the plotter icon located in the lower right corner of the AutoCAD window.

When the plot is complete, a notification bubble will appear.

 If you do not want this bubble to appear you may turn it off. Right click on the plotter icon and select "**Enable Balloon Notification**". This will remove the check mark and turn the notification off. You may turn it back on using the same process.

When you click on the "Click to view plot and publish details..." , on the bubble, you will get a report like the one shown below. The report will list details about all of the drawings plotted in the current drawing session.

 You may turn Background Plotting ON or OFF. The default setting is OFF.
To turn it ON or OFF:
select **Tools / Options / Plot and Publish tab.**

Notes:

APPENDIX E
METRIC CONVERSION FACTORS

Multiply

Length	By	To Obtain
centimeter	0.03280840	foot
centimeter	0.3937008	inch
foot	0.3048a	meter (m)
foot	30.48a	centimeter (cm)
foot	304.8a	millimeter (mm)
inch	0.0254a	meter (m)
inch	2.54a	centimeter (cm)
inch	25.4a	millimeter (mm)
kilometer	0.6213712	mile [U.S. statute]
meter	39.37008	inch
meter	0.5468066	fathom
meter	3.280840	foot
meter	0.1988388	rod
meter	1.093613	yard
meter	0.0006213712	mile [U. S. statute]
microinch	0.0254a	micrometer [micron] (mm)
micrometer [micron]	39.37008	microinch
mile [U.S. statute]	1609.344a	meter (m)
mile [U. S. statute]	1.609344a	kilometer (km)
millimeter	0.003280840	foot
millimeter	0.03937008	inch

Multiply Length	By	To Obtain
rod	5.0292a	meter (m)
yard	0.9144a	meter (m)

Area

acre	4046.856	meter2 (m2)
acre	0.4046856	hectare
centimeter2	0.1550003	inch2
centimeter2	0.001076391	foot2
foot2	0.09290304a	meter2 (m2)
foot2	929.0304a	centimeter2 (cm2)
foot2	92,903.04a	millimeter2 (mm2)
hectare	2.471054	acre
inch2	645.16a	millimeter2 (mm2)
inch2	6.4516a	centimeter2 (cm2)
inch2	0.00064516a	meter2 (m2)
meter2	1550.003	inch2
meter2	10.763910	foot2
meter2	1.195990	yard2
meter2	0.0002471054	acre
mile2	2.5900	kilometer2
millimeter2	0.00001076391	foot2
millimeter2	0.001550003	inch2
yard2	0.8361274	meter2 (m2)
fathom	1.8288	meter (m)

Symbols of SI units, multiples and sub-multiples are given in parentheses in the right-hand column.

Appendix E-2

APPENDIX F
DRAWING SCALES

Scale	Drawing Scale Factor	Adjusted Scale times Paperspace
1/16=1'	192	1/192xp
3/32=1'	128	1/128xp
1/8=1'	96	1/96xp
3/16=1'	64	1/64xp
1/4=1'	48	1/48xp
3/8=1'	32	1/32xp
1/2=1'	24	1/24xp
3/4=1'	16	1/16xp
1=1'	12	1/12xp
1-1/2=1'	8	1/8xp
3=1'	4	1/4xp
1=10'	120	1/120xp
1=20'	240	1/240xp
1=25'	300	1/300xp
1=30'	360	1/360xp
1=40'	480	1/480xp
1=50'	600	1/600xp
1=60'	720	1/720xp
1=80'	960	1/960xp
1=100'	1200	1/1200xp
1=200'	2400	1/2400xp
1=10	10	1/10xp
1=20	20	1/20xp
1=16	16	1/16xp
1=30	30	1/30xp
1=40	40	1/40xp
1=50	50	1/50xp
1=100	100	1/100xp
2=1	0.50	2xp
4=1	0.25	4xp
8=1	0.125	8xp
10=1	0.10	10xp
100=1	0.01	100xp

NOTES:

INDEX

NOTES:

This page blank intentionally.

NOTES:

This page blank intentionally.

AutoCAD 2007 trial software

The enclosed CD contains a fully functioning 30-day trial version of AutoCAD 2007 software. This software has been provided to give you a preview of how easily AutoCAD 2007 installs and is used in conjunction with this Exercise Workbook. After 30 days, from the day you install the software, you will not be able to open the program. At that time you will have the opportunity to activate the software by purchasing the actual software license.

IMPORTANT: The 30-day trial version of AutoCAD can be loaded only once on a single computer. This applies to all 30-day trials regardless of where you obtained them, including from other Exercise Workbooks by Cheryl Shrock.

For example, the 30-day trial from the Advanced Workbook will not install on a computer on which you previously installed the 30-day trial from the Beginning Workbook.

If you are attending a school, check with your instructor for educational discounts.
Or visit: www.autodesk.com for additional purchasing information.

AutoCAD® 2007 software enables highly efficient creation of a single drawing as well as timely coordination of drawing sets. Everyday tools like the table object and tool palettes boost productivity, and the new Sheet Set Manager feature permits content control across entire sets of related drawings. Sheet sets can be composed and collated in a single view, then shared with the entire project team using plots, eTransmit, or DWF™ (Design Web Format™) files.

AutoCAD 2007 system requirements
Refer to page 1-6

How to install AutoCAD:

Note: AutoCAD 2007 can coexist with earlier versions of AutoCAD and AutoCAD LT.

Two disks are required to install AutoCAD. Insert AutoCAD 2007 Disk 1 of 2 to start the installation. When prompted, insert AutoCAD 2007 Disk 2 of 2 to complete the installation.

1. Insert the AutoCAD 2007 Disk 1 of 2 into your computer's CD-ROM drive.

2. In the AutoCAD Media Browser, on the install tab, click Stand-Alone Installation.

3. Under Install AutoCAD 2007, click Install.

4. On the Autodesk Setup page, click OK to install required support components.

5. In the AutoCAD Installation wizard, follow the directions on each page. When prompted, insert AutoCAD 2007 Disk 2 of 2 to complete the installation.

6. When the installation is complete you will be prompted to:
   "Authorize AutoCAD" or "run AutoCAD without authorizing."

   Select "**run AutoCAD without Authorizing**"

Note: This trial software will stop working 30 days from the date of installation.